# DIVING FOR SCIENCE

The Story of the Deep Submersible

By the same author

*Exploring the Ocean Depths:* The Story of the
Cousteau Diving Saucer in the Pacific

EDWARD H. SHENTON

# DIVING
# FOR SCIENCE

The Story of
the Deep Submersible

W · W · NORTON & CO · INC ·
NEW YORK

Copyright © 1972 by W. W. Norton & Company, Inc.

FIRST EDITION

Library of Congress Cataloging in Publication Data
Shenton, Edward H.
    Diving for science.

    Bibliography: p.
    1. Submarine boats.  2. Oceanographic research.
I.  Title.
VM365.S42        551.4'6'0028        74–90990
ISBN 0–393–06380–1

Published simultaneously in Canada
by George J. McLeod Limited, Toronto

PRINTED IN THE UNITED STATES OF AMERICA

1   2   3   4   5   6   7   8   9   0

To the Pioneers of the Deep Submersible who foresaw the need to develop these vehicles.

# Contents

*Photographs may be found following p. 128.*

# Preface

The story and series of events leading to the development of the modern research submersible began shortly after man became interested in the sea many centuries ago. The actual research submersible has come into being only since 1960 —a period of intensive development. I have tried to assemble some of the more interesting and significant events that highlight the evolution of the manned submersible as it grew out of the submarine.

One of the reasons for writing this book has been to draw together both the fascinating history of submersibles and the reason for their existence in terms that anyone can understand. In doing so, I felt it necessary to explain in a relatively simple way a little of submersible design and function—how they are built and why they function as they do—since I believe this is important in understanding their ability to do work underwater. Another reason for writing the book is that the subject of manned submersibles and the accomplishment of men underseas is an exciting field that has received relatively little attention and deserves to be clearly explained to the general reader.

During the time I was gathering information on the activity of various vehicles, I received a great deal of assistance from several individuals to whom I am most grateful. Mr. John A. Pritzlaff of Westinghouse Ocean Research and Engineering Center, Annapolis, Maryland, was especially helpful in a critical review of my manuscript and contributed substantially to the factual content of several chapters. I wish to thank Lieutenant Commander Donald Blake of Submarine Development Group One, San Diego, for providing information on the submersibles operated by SUBDEVGRU One and for several United States Navy photographs of these vehicles. Joe Thompson, who worked with me on the Westinghouse/Cousteau *Diving Saucer* Operation in 1965, and who later became Chief Pilot of *Beaver IV* at North American Rockwell, kindly supplied information and a number of photographs. I am especially grateful to all the many companies and groups too numerous to mention separately for their cooperation in providing me with details of their vehicles for inclusion here.

The drawings throughout the book were done by my wife, Karyl, who donated the time, while more important items of business had to wait. Finally, I was assisted by three indispensable secretaries, Peggy Martin, Brenda Kuetzing and Jan Heiges, in typing and assembling the final manuscript.

La Jolla, California                        Ned Shenton
October 1971

# DIVING FOR SCIENCE

The Story of the Deep Submersible

# 1

# A Fish at the Bottom

While France was making last minute preparations for exploding its first atomic bomb in the Sahara, and President Eisenhower was continuing to de-emphasize the importance of the scientific exploration of outer space, a small team of U.S. Navy specialists were preparing a destroyer and tugboat tender for an unusual mission. After steaming to a site 200 miles south of Guam, two of the men debarked and slid down a 13-foot tower about 25 inches wide and wriggled into a sphere 6 feet in diameter. Racks of instruments lined its interior. Two small seats, kneeling cushions, and oxygen bottles were crowded into the small space. It was the day of the Grand Plongée.

For two years the men had been planning the Big Dive into the deepest part of the Pacific Ocean. They were Jacques Piccard, son of the inventor Auguste Piccard, who constructed the bathyscaphe *Trieste;* and Lt. Donald Walsh, Officer in Charge of the Navy's Project Nekton. Walsh's ambitious project was an attempt to dive more than 35,000 feet down into the Challenger

Deep in the Marianas Trench of the Pacific Ocean. This was dive number seventy for the seven-year-old Italian bathyscaphe. The 54-foot long submersible belonging to the Navy Electronics Laboratory in San Diego had been brought to Guam by ship and towed 200 miles to the world's deepest trench.

The dive began at 0823 as *Trieste* began her slow descent to the bottom. Forty-five minutes later they had only reached 800 feet as the gasoline in the float cooled slowly. At 1,000 feet only a faint trace of light filtering down from the surface was perceptible. Minute specks of plankton in the water moved past the view port as the bathyscaphe began to pick up speed. Soon the craft was plunging at 3 feet per second—about the maximum descent rate for the 150-ton vehicle. By the time 3,000 feet was reached, the *Trieste* had been sinking for nearly an hour. They were passing one of the depth records held for many years by Prof. William Beebe and Otis Barton, made in 1934 by Beebe's *Bathysphere*. At 0928 a slight leak was discovered in one of the hull connectors. Walsh watched it closely. Usually these leaks would seal as the boat went deeper. At 4,500 feet they passed Barton's deepest dive in his *Benthoscope*, a dive made in 1949 off California.

Above the plunging *Trieste*, on a rough and choppy sea, rode the tending ship *Wandank*, which had towed *Trieste* to the trench, and *Lewis*, the U.S. Navy destroyer. Walsh maintained communication with these vessels using the wireless underwater telephone called a UQC by the Navy.

By ten o'clock the men had reached almost 10,000 feet. The cramped cabin in which Walsh and Piccard sat became chilly as the water outside grew colder. The trickle leak stopped as suddenly as it had started. Nothing was visible from the port; blackness pressed against the 7-inch-thick plexiglas. Passing 10,500 feet, the craft and crew went by another record, made by the French navy's bathyscaphe *FNRS-3* in a dive off Japan in 1958. At 1100 *Trieste* overtook its own record dive of 18,500 feet made

16

a few months earlier; and shortly after that, the 24,000-foot dive of January 8. Pilot Piccard and Pilot Walsh were finally in virgin waters.

The following log of the remainder of the dive, taken from Walsh's tape, relates the experience of the *Trieste* divers:

"Time 1206. We are passing 32,400 feet. A sharp explosive noise sounded outside, rocking the whole bathyscaphe. We are trying to locate it. Jacques thinks it was one of the incandescent lights outside the port, but that doesn't seem to be it. Nothing else is affected and we continue to descend. All systems are normal.

"Time 1238. Six thousand fathoms. No indication of the bottom so far on the fathometer. Jacques has the bow light on and is periodically looking out, but there's no sign that we're getting close. We're moving slowly now because we've been dropping shot to slow our descent speed.

"Time 1258. We have the bottom in sight on the fathometer at 42 fathoms. Thirty-two fathoms. Fairly steep curve or slope here, 28, 25 fathoms, 24—now we're getting a nice trace. Twenty-two, still going down. Twenty fathoms. Looks like we finally found it, Jacques. Sixteen fathoms, 14 fathoms. When we reach 10 fathoms we can start looking, can't we? Quite clear out there. Thirteen fathoms, 12 fathoms—72 feet. Ten fathoms, making a nice trace now. Going right down. Nine fathoms, 50 feet. Any time now. Still going down, 8 fathoms, 7½; nice trace there. Should be a very gentle landing the way that curve is. Going very slowly. May stop before we get to the bottom. About 5 fathoms—30 feet. You say you saw some little animals, Jacques? He says they're Medusa, or maybe red shrimp, about an inch in diameter. Four fathoms, 24 feet. Should see the bottom any minute now. Quite light outside with the vehicle's light reflecting off the bottom. Two fathoms—just about to arrive on the bottom. We've landed! Fish —a fish about a foot long—at the bottom of the ocean! What does it look like, Jacques? He says it's like a sole—it *is* a sole. Does he

17

have eyes or does he appear to be blind? Is he in the sand? Did he move away or is he still there?

"We're off bottom. I think I'll turn off the fathometer here, Jacques. Check signals to the surface. Time 1306. Landed on the bottom, at 35,800 feet!"

The first day of the space age is usually taken to be October 4, 1957, when the Russians placed the first artificial earth satellite in orbit. Far less widely remembered is the date of January 23, 1960, which is considered by many ocean scientists the beginning of the submersible age—the day *Trieste* made her dive to the bottom of the Challenger Deep. It happened with a minimum of fanfare and publicity and probably passed unnoticed by a busy world.

# 2

# The Crude Beginnings

## 4th Century B.C. to 1895

There is much to be said for the statement that little if anything is ever invented . . . it's just rediscovered after being forgotten for a period of years or centuries. Thus, many of our new and shiny devices are merely retreads or things correctly engineered that were conceived several hundred years ago. As the working details of the aerospace age were imperfectly conceived in the seventeenth century, so too were some of the ideas for underwater technology. Primitive versions of submarines began to appear about that time, and it's remarkable that Mother Shipton, the 16th-century seer, had the incredible foresight to make the far-reaching statement: "Under Water men shall walk, Shall ride, Shall sleep, Shall talk."

This rather amazing prediction made four hundred years ago covers much of the new and exciting activities going on or planned for the undersea age about to involve us. In between is a long series of events and progressive developments that have made the prediction a reality.

Man had ventured underwater in bells and diving rigs and as a free diver long before the first diving boat or submarine. Reliable accounts describe the heroic feats of divers before the time of Christ in warring attacks on port facilities in Greece. The use of crude animal-skin breathing devices was attempted as early as the fourth century A.D. Apparently in the late fifteenth century a hand-propelled wooden submarine was built by Roberto Valtino of Venice, but no record of its use remains. Also at that time, Leonardo da Vinci was working with concepts for submarine boats along with numerous underwater warfare techniques. Da Vinci felt strongly about the possible unrestricted use of the submarine, so much so that he would not divulge the nature of his plans for an undersea vehicle. In his notebook he wrote:

> How by an appliance many are able to remain for some time under water. How and why I do not describe my method of remaining under water for as long a time as I can remain without food; and this I do not publish or divulge on account of the evil nature of men who would practice assassinations at the bottom of the seas, by breaking the ships in their lowest parts and sinking them together with the crews that are in them; and although I will furnish particulars of others, they are such as are not dangerous, for above the surface of the water emerges the mouth of the tube by which they draw in breath, supported upon wineskins or pieces of cork.

While da Vinci and others of this period only envisioned underwater vehicles, it was not until about 1620 that a man from Holland, Cornelius van Drebbel, constructed a real submarine. He built two boats for use on the Thames. The larger one was designed for twelve oarsmen and as many passengers, who entered through leather "hatches" or joints which were watertight. The hull was built of wood, leather covered, and a coat of grease was smeared over the vehicle. The structural strength was gained from iron bands.

Although various accounts disagree, van Drebbel did submerge the vessel for short distances in the Thames. According to one such account, the submerged craft took passengers between Westminster and Greenwich over a period of several hours. While it may be difficult to substantiate this claim, it is of interest that the van Drebbel craft contained a working life support system. He described a mysterious chemical he called "quintaessentia" which revived the atmosphere inside the boat. Sir Robert Davis, the well-known British chronicler of diving apparatus, believes this may have been a soda ash that absorbed the carbon dioxide. No original records or drawings have survived for the boat although one account alleged that the length was about 270 feet.

Between this first documented operable vehicle and the next significant step there were half a dozen attempts at submarines —none of any particular success or ingenuity. Progress was slow simply because the materials available were totally inadequate for underwater work. Leather was the only material for sealing out the water; there was no protection against the biting cold; life support as we know it was nonexistent for diver's dress, and there was no propulsion for the vehicles except cranks, pedals, and oars. Nevertheless, underwater activities struggled on under these crude conditions. Sir Edmund Halley, best remembered as an English astronomer and the discoverer of Halley's Comet, designed and constructed a diving bell in 1692. While not a direct predecessor of the submersible boat, the Halley diving bell showed very sound concepts and details that make it the forerunner of the modern bell. From each of the realized endeavors there is bound to be a cross-fertilization and generating of enthusiasm. Halley set a record by placing four men, including himself, on the bottom in divers' dress at 60 feet for ninety minutes. Air was supplied in barrels lowered to the base of the bell and then pumped by a tube into the bell.

While the Halley bell was an enormous improvement over anything previous, it lacked mobility and needed a large ship and crew to handle it. Further, the rig was expensive to operate.

Of the several attempts to advance underwater vehicles, one failure is amusingly related by the late James Dugan, an outstanding historian of undersea activities. A dockyard laborer and mechanic named John Day, who worked at Yarmouth, England, built a small vessel, a "tub," in 1772 in which he reportedly descended 30 feet for some twelve hours. A friend, Dr. H. D. Falck, wrote of Mr. Day, "He was a man very illiterate . . . and his temper was gloomy, reserved, and peevish, his disposition penurious . . . and unshaken in his resolution." After the first submergence, Day found capital in London to back a deeper venture using a 50-ton sloop. In 1774, Day, his converted sloop-submarine now painted bright red, announced to the world that he was going to dive to 300 feet for twenty-four hours! The confident but doomed man entered his vehicle provisioned with biscuits, a candle, and a clock and was neatly sealed in by Dr. Falck. The submarine carried 30 tons of stone ballast, but even in the ballasted condition, the ship would not sink. Day ordered more stone placed in the hold. Dr. Falck wrote, "She was sunk at two o'clock in the afternoon of June 20th and Mr. Day with her into perpetual night." Unfortunately, Mr. Day had not provided a suitable ballast control or variable ballast and the good ship *Maria* and Mr. Day were never seen again, although rescue attempts continued for several months. Dr. Falck concluded in his account that John Day must surely have frozen to death. Far more likely, he drowned as the ship collapsed under pressure.

The perils and pitfalls of the early experimenters were numerous and seem elementary from our vantage point of progress in technology. The heroes of the submarine were bold and brave, just as the experimenters of the airplane era. Flying blossomed in a relatively short period of a quarter century, but submarining struggled over two centuries before achieving its goal.

While John Day sank ignominiously, one of the significant milestones in the development of the military vehicle, the submarine, was achieved by David Bushnell in 1773—it was his famous *American Turtle* built as an undersea attack vessel during the American Revolution. Bushnell, a Yale graduate, proposed to build a vehicle with which he could destroy British ships. The name *Turtle* was derived from the fact that the wood and iron-reinforced halfshells when joined looked like turtle shells. At the top was a conning tower with room for the operator's head with a viewing window above water. The rest of the *Turtle* was submerged in a vertical position. Water ballast was admitted by a valve into two separate tanks below and could be discharged by

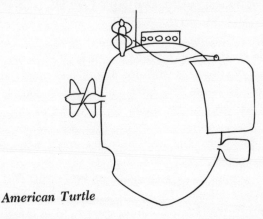

*American Turtle*

a foot pump—a more practical arrangement than poor Mr. Day had foreseen. Propulsion consisted of a hand-driven spiral screw (the first known application of such a device) and a vertical screw to submerge the vehicle once ballasted. Enough air was present for a half-hour submerged mission as well as a two-tube ventilator when the conning tower was awash. Navigational instruments consisted of a phosphor-coated compass and a glass pressure gauge. The speed of the vessel was reported to be three knots. Its armament was referred to as a "torpedo"; it resembled a mine with 150 pounds of powder in a container and was mounted on the stern of the *Turtle*. Bushnell's objective was to maneuver the *Turtle* under cover of darkness up to a British man-of-war, submerge under the ship, and place the torpedo by screwing it into the ship's bottom. A clockwork fuse allowed a half hour for the *Turtle* to escape.

General Washington was convinced that he should risk the *Turtle* against the British in 1776 since America at the beginning of the Revolution had no warships and was badly overpowered by the British. Thus, the submarine became the desperate hope of the weaker power as it always did in the early days of submarines.

A brave army sergeant named Ezra Lee was chosen to pilot the craft and was trained to operate the *Turtle*. The HMS *Eagle*, Admiral Howe's flagship, was selected as the target for maximum effect on the enemy. Proper timing on drifting with the tide and several hours of strenuous work for Lee on the rudder and hand screw got *Turtle* to the ship after midnight. Cautiously, he submerged over 10 feet to get under the big 64-gun ship, only to find the bottom sheathed with an impenetrable copper sheathing for worm protection. The torpedo screw could not attach the powder charge. Lee, undaunted, tried for several hours to find a spot without copper. Finally, he gave up and in the first light of dawn, narrowly escaped being captured by patrolling British ships. Several further torpedo forays were tried in the New York area—all without success as the British were by then alerted to the danger. Although the submersible *Turtle* performed admirably, it was never able to complete its mission. Bushnell continued his work for the next twenty years and finally retired in disgust without ever bringing his invention to national fame. Had events worked differently at any of the several critical moments when a ship was to be blown up, the eventual development of the military submarine might have been achieved a century sooner.

The next development in the long saga of the submarine is the story of Robert Fulton who, like Halley, is best known for his endeavors in other fields. Fulton was an inventor from Pennsylvania who is remembered by most schoolchildren as the father of the steamboat. He migrated to France via England in 1800 in search of a better opportunity for himself. In Paris he ran into Bushnell, who was then peddling submarine designs to the French Government. Fulton eagerly competed with Bushnell by building a 21-foot torpedolike vessel of wood. He called this sub *Nautilus*, the first of many underwater boats to follow with this name.

By 1801, Fulton managed, with the backing of two famous French scientists, Monge and Laplace, to bring the torpedo-toting sub to the attention of the French Ministry of Marine and

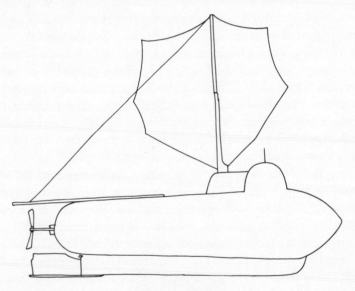

Fulton's *Nautilus*

ultimately Napoleon. Fulton designed *Nautilus* with a hollow cast-iron keel that was flooded to sink and pumped out to rise. She sailed with the observation dome awash and a strange, large folding sail protruded when she was surfaced. Two sailors hand-powered her by cranking a bladed propeller—an improvement over the Bushnell spiral screw. *Nautilus* carried a gun powder mine to be driven into a ship with a spike.

During a special visit with Napoleon, Fulton proposed immediately using the *Nautilus* against the British. Napoleon was interested but cool and Fulton became impatient at the delay. Monge and Laplace controlled his temper and arranged a demonstration before the Ministry of Marine in which the *Nautilus* attacked a small sloop from a distance of 650 feet while completely submerged. Before the horrified eyes of the viewing admirals, the 40-foot ship was torn to bits, and a 100-foot column of water was raised. Everyone was convinced that the submarine could destroy a warship—everyone but Napoleon, who had more grand ideas than a silly little underwater boat. Fulton felt that his demonstration was adequate and could not wait any longer. He took his ideas and drawings to the British Admiralty in 1804 and allied himself to William Pitt who gave him minor support to pursue the development of floating mines. The building of another *Nautilus* was delayed just long enough for Lord Nelson to destroy the French and Spanish fleets at Trafalgar, eliminating the pressing need for Britain to have a secret weapon.

Robert Fulton—a frustrated submarine builder—sailed for his home in America in 1806. He had complete plans for a 35-foot, 30-mine submarine with many innovations. He submitted these plans to President Jefferson, who was not interested enough to reply. In desperation, Fulton with his merchant friend, Robert Livingston, designed and built the steamboat *Clermont*. Ironically, he believed that the steamboat was not progress but retrogression. His was the seventh steamboat to come along, but it came at the right time—when the demand and the public awareness was right. Anybody could build a steamboat, Fulton believed, but what the world needed was a revolutionary approach —an undersea boat! In his last years, Fulton was famous and wealthy, but he was driven by his belief in the submarine. He built the *Mute* in 1815. She was 80 feet long, and required a crew of a hundred men to crank the propeller. However, Fulton died and the *Mute* was never launched. If, as with the *Turtle* and David Bushnell, Fulton had been able to pursue the submarine to more success, there might have been realities like Jules Verne's imaginary *Nautilus* in the mid-nineteenth century.

After Fulton came the Bavarian, Wilhelm Bauer, with several vehicles in 1850. Bauer's boats, while considered successful, produced no successor or significant developments. He did, however, use a new technique of submersion which reappeared some one hundred years later. Instead of bow planes for diving, a heavy weight was moved forward affecting the bow-down angle and forcing a kind of erratic diving and pitch instability.

The submarine was being regarded more and more as a military weapon, and a series of boats known as the David boats, referring to the Biblical hero, were used by the Confederate forces during the Civil War to attack blockade ships of the North. The Davids were about 50 feet in length, very narrow, and were manned by a crew of nine—eight of them who put their shoulders to turning the single shaft propeller. There was no life support system—which unfortunately was responsible for the demise of several crews by suffocation during experimental dives. It was in these vehicles that the miner's $CO_2$ meter, a canary, was ably

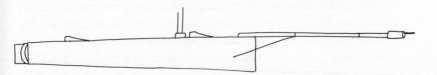

*David of Hunley*

employed to warn of oxygen exhaustion. The one David boat that became an historical first was called *David of Hunley,* or just *Hunley,* for her designer Captain Hunley. Actually this vessel was smaller than the earlier boats, but she carried a 10-foot spar torpedo. After a number of mishaps, the *Hunley* was sent to attack the iron-clad Union ship *Housatonic*. The submarine sneaked up in the dark, placed her charge against the 13-gun corvette, and the gallant Lieutenant Doon, commander of the *Hunley*, pulled the firing lanyard. In a blinding flash, the *Hunley* broke up from the concussion and immediately sank stern-first with all hands. Shortly after, the *Housatonic* also sank. It was significant as the first naval encounter in which a submarine sank an "enemy" ship—although another seventy-five years passed before an American submarine sank a ship of another country in the Second World War.

By 1869, at least twenty-five authenticated submarines were built and had dived with varying success. This was also the year in which Jules Verne published his prophetic novel, *Twenty Thousand Leagues Under the Sea*, which projected well into the twentieth century with his *Nautilus* and the amazing Captain Nemo.

After the end of the American Civil War, there was a noticeable increase of variety in submarines and submersible boats. There were as many as twenty large submarines built in the last part of the nineteenth century; few of them were successful. Electric, steam, and diesel power were all on the horizon, signaling a breakthrough in propulsion, speed, and endurance.

One of the few remaining relics from this era is on exhibit at the Navy Shipyard in Washington. It is called the *Intelligent Whale* and was a handcranked experimental submarine built from 1864 to 1872. It was the first official U.S. Navy attempt at a submarine. The *Intelligent Whale* was designed by Skovol S. Merriam, who made an agreement in 1863 with August Price and Cornelius S. Bristol to build a boat from plans of his invention. Price and Bristol were to furnish the capital in the amount of $15,000. In April, 1864, the American Submarine Company was formed and took over the interests of Price and Bristol. Years of

28

litigation followed, during which the boat was finally sold at a sheriff's sale. When title was established by the court on October 29, 1869, the vessel was sold to the U.S. Navy. About three years later the first trial was made, but it was unsuccessful—reportedly thirty-nine persons drowned. The Secretary of the Navy re-

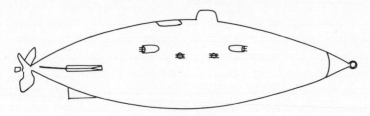

*Intelligent Whale*

fused further expenditures and the project was abandoned. The boat probably would have lain untried had it not been for General Sweeney, who persuaded two others to venture in it with him. They dived to 16 feet and then General Sweeney climbed into the diver's suit and passed through the manhole or hatch in her bottom. He then placed a torpedo under a scow, re-entered the boat, and exploded the torpedo by a lanyard. The scow was blown to pieces. The exact procedure used by General Sweeney to get out and then back inside the boat from the bottom was not described. The after compartments were filled to sink the boat by opening a valve; to raise it, the water was ejected by a pump or forced out by compressed air. It is estimated that the boat could stay down about ten hours. Thirteen persons could be accommodated aboard her; however only six were needed to make her operational. Her speed was about 4 knots by the use of a handcrank. The *Intelligent Whale* was 30 feet long and about 9 feet in depth. According to the *Army and Navy Journal* of September 14, 1892 she was known in the Navy Yard as the "Intelligent Elephant". Although the vehicle saw little action, the fact that a "swim out" hatch was used is of interest since it precedes the better known example of a bottom hatch built by Simon Lake twenty years later.

Thus the crude beginnings of the submersible gave way to a period in which the developments would have significant bearing on all subsequent underwater craft.

# 3

# The Founding Fathers

## 1865-1954

Many experimental submarines were built in Britain, France, Germany, Spain, Sweden, and America, during this period, and two men are probably the best known and contributed more to the modern submarine and the research submersible. One, an Irishman, John P. Holland, emigrated to America in 1873 with a vision of a military submarine that was to engage him in persistent endeavor for nearly thirty-five years trying to sell the concept to the U.S. Navy. The other, Simon Lake, an American, grew up in the 1880s with a similar dream after having read Jules Verne's tale of the *Nautilus*—his aim was more toward exploration and salvage. These two men are the fathers of the two types of underwater vehicles. Holland truly laid all the conceptual ground work for modern warfare by nuclear submarines. His original hull designs in 1875 were finally adopted by the nuclear boats seventy-five years later when submersibles departed from the old "U" boat design which was more suited to surface running than submerged speed. Lake, on the other hand, designed and built several submersibles along the lines of our present research vehicles, and developed such innovations as a swim-out lock, forward view ports, and wheels to ride on the bottom.

John Holland had become a school teacher in Paterson, New Jersey. A friend urged him to submit his design for a submarine boat that he had prepared in Ireland fifteen years earlier. With slight revisions, Holland sent his plans to the U.S. Navy, only to be rebuffed. As most of the rest of the nineteenth-century inventors did, Holland, with the help of a backer, went ahead to build his submarine himself. He started by making a scale model and from this he drew up his detailed plans and proceeded in 1878 to build the *Holland No. 1*. This prototype, 14 feet long and 3 feet in beam, gave Holland the chance to learn more about the ways to build submarines. *No. 1* had many shortcomings— mostly in the Brayton gasoline propulsion motor which never functioned. The launching of *No. 1* was viewed by many curious spectators in Paterson who knew nothing of the wonder of a submarine. They were amused to watch the small vessel slip into the waters of the Upper Passaic and in a few minutes settle quickly to the bottom. Most popular accounts still attribute this

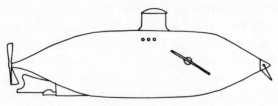

*Holland No. 1*

mishap to someone forgetting to put in the drain plugs. Charles Morris, in his biography of Holland, points out that to the contrary it was faulty riveting that failed. The boat was raised the next day and dried out. *No. 1* did make a successful dive without the power of the contrary Brayton engine and Holland remained submerged for over an hour. He had gained as much information as he felt he could, so rather than modify the *No. 1*, he removed salvagable things such as the propulsion machinery and scuttled the hull. Fifty years later this hull was located on the river bottom, salvaged, and placed in the Paterson Museum where it is today.

Holland received considerable financial backing from Irish militants—the Fenian Brotherhood—who wished to sink British shipping. Holland in Irish style would also have enjoyed harassing the British. The Fenians secretly provided funds to Holland and *Holland No. 2* materialized. But in spite of the secret efforts, the press had been aware of what was going on and dubbed it *The Fenian Ram.* The name stuck. The *Ram* was 30 feet long with a 6-foot beam and displaced 19 tons. The $1\frac{1}{16}$-inch steel plate allowed a calculated 50-ton ramming force. Holland had encouraged the Brayton engine manufacturer to eliminate the faults of the unit in *No. 1*, and this they did with a 17-HP engine with a surface speed of 9 mph. Holland staged trials off Brooklyn in 1883 with a submerged trial of 21½ hours and finally a dive to 60 feet.

The strange little man in the bowler hat became known for his submarine boats, but he was nearly destitute. He had been working for no salary in the hopes he could sell his invention to the Fenians. In late 1883, much to his surprise, the hand that had fed him—after a fashion—was turned against him. One night his Fenian partners and Irish rebels stole the *Ram* as well as another prototype—a 16-foot scale model of the *Ram.* Both boats were towed up the East River, but the smaller boat was lost when it flooded through an open hatch and sank in 110 feet of water. Holland never understood the theft, and he never went after the *Ram*, which wound up in a barn in Connecticut until 1927. Then it was placed in a park, the West Side Park at Paterson, New Jersey, only a short distance from the river where Holland had launched *No. 1*.

Holland was disappointed in losing the financial support of the Fenians but he did not slow down, nor did he veer from his conviction that another boat should be built. Also, by this time, John P. Holland had become an American citizen, and he was less interested in the Irish cause that had inspired the building of the *Ram*.

The fourth of Holland's creations was not fostered by naval interests but by a young Army artillery officer and ordnance expert who like Holland had taught school (MIT), and who had a great desire to see an underwater boat with a pneumatic dynamite gun. This man was Lieutenant Edmund Zalinski of the U.S. Fifth Artillery and founder of the Nautilus Submarine Boat Company. Zalinski looked on the boat as "a floating gun carriage," and in 1884 he hired Holland to design and oversee the building of the submarine. Holland took this opportunity to further some of his experimental designs that had been stopped by the breakup with the Fenians. He had realized from his first diving ventures the necessity of being able to navigate underwater and also see out. He hoped to make improvements on a periscope—or more accurately a "camera obscura"—and had ideas on automatic steering as well as new ballast arrangements. This design called for a boat 50 feet long, 8 feet in diameter, to be built of wood on iron frames. But by 1885, the financial fortunes of the Nautilus Company did not allow Holland the latitude he needed and he felt the boat was not going to be an improvement. The launching proved disastrous. The ways were poorly constructed and as the heavy boat slid down toward the water, the structure collapsed, severely damaging the sub. Although Zalinski insisted that it be raised and repaired and several dives were completed, the project was abandoned.

Holland was still sure of his design principles for a submarine, even though most of his competitors insisted on building boats in a different manner. He strongly believed in a boat with positive buoyancy that must drive itself down underway with planes and a bow-down angle. The other method of submergence that was gaining support was the "even keel" belief that a vehicle should submerge horizontally. People skeptical of Holland's design believed angles of 15° to be dangerous and unstable. Holland realized, however, that if his boat were to be involved in warfare, even-keel descent was too slow, while a powered dive could get the submarine under quickly.

In 1888, the U.S. Navy had advertised for open competition for the design of a submarine. Holland in partnership with a shipyard won the competiton but was disqualified on a technicality. The design was rebid, and again Holland won only to have President Harrison redirect the funds for a surface ship instead. Holland was worn out and down to his last penny, so for several years he worked for his friend, Charles Morris, on the design of a flying machine.

By 1890 Holland had conceived a far better idea for building a submarine that would be a successful weapon of war than most of his Navy contemporaries and other designers working in America. His guiding principle was that a sub must be independent of the surface to be effective. Yet the Navy, he said, "wished for a deck on which to strut and enjoy sunshine." Holland foresaw the ultimate breakthrough that nuclear power was to bring the submarine sixty years later by insisting that a military submarine had to be self-sustaining and remain submerged.

By 1893, another backer became interested in the submarine, one who knew the Washington political gauntlet that must be run to carry off the prize. E. B. Frost, with Holland as general manager, formed the John P. Holland Torpedo Boat Company. With real perseverance, a new design, and some new political friends, Holland was given a government contract of $150,000 to build a steam submarine in 1895—roughly thirty-five years after he had first designed a practical diving boat. It may seem after all this endeavor that he had achieved his goal, but not so. The Navy chose to play a strong part in the design. Before long Holland was having many disputes with the Navy experts over the basic design and realized that his fifth boat, which the Navy called *Plunger*, was going to be a dismal failure. The steam plant looked as though it would generate so much heat that the crew would be roasted inside—like being in a pressure cooker. Over a year or so the boat had grown like Topsy and lost all connection it had with the original design. Holland felt helpless to fight the Navy "brass." Instead, true to the spirit of an inventor who wishes above all else to see his idea succeed, John Holland convinced his company management that they must back him in building the sixth Holland boat as a private venture. The company, also foreseeing the failure of the *Plunger*, agreed, and the *Holland VI*, the culmination of John Holland's efforts, was launched in mid-1897 while the *Plunger* was still under construction. The vessel was 53 feet long, beautifully shaped with a hydrodynamic hull, propelled by a fully workable gasoline engine, and capable of meeting all of the Navy's requirements. After extensive trials and demonstrations, the *Holland VI* was purchased by the U.S. Navy in April, 1900. This submarine, rather than the *Plunger*, which did fail, became the first commissioned U.S. Navy submarine and went on to dictate the specifications of the first class of six subs built by the Holland Company which had by then changed its name to the Electric Boat Company. Unfortunately John Holland was squeezed out of his company and lost all the patent rights. In 1914 he died penniless and barely known for the significance of his achievement. Few, if any, of the military experts

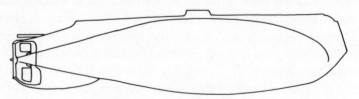

*Holland VI* or U.S.S. *Holland*

realized the future role of the submarine. No one could foresee that the basic, short cigar shape of the *Holland VI*, also known as the U.S.S. *Holland*, would be almost the same as the ultimate shape of the high speed *Albacore* which in 1959 led to the diesel SS580 BARBEL class and the nuclear SSN 585 SKIPJACK class. So the Navy in the next few years wasted no time in changing this shape of Holland's design to the long narrow boats with decks "on which to strut." Holland's integrated joy stick control in which one pilot commanded the sub in *Holland VI* was forgotten for nearly sixty years until the high speed nuclear boats used this same type of one-man control. One wonders what might have happened had John P. Holland not had such a struggle with the government and the U.S. Navy over his boats.

While Holland was in the midst of his fourth submarine in about 1885, other quite varied submarines of all shapes and sizes were being built. Few of them showed any similarities in design yet several employed electric principles. The British *Nautilus* designed by Campbell and Ash used a variable ballast system for ascent-descent as well as minor adjustments. It was composed of eight large pistons approximately 3 feet in diameter placed along the side of the boat. When these were retracted into the hull, the displacement was changed, the boat lost buoyancy and could descend. A variation of this technique for a variable ballast system is employed in several of our present submersibles. Several of the larger vehicles were extremely long and narrow and used a water ballast system. Unfortunately, many, like the *Nordenfelt*, a Swedish design of Cyrus Nordenfelt, were highly unstable in the pitch plane because the water in the ballast tanks and boilers would surge uncontrollably, causing wild dive angles. Although several of these boats had ingenious propulsion systems, none was perfected to the point of being considered successful —or rather none really led to a future generation of submarines in the way that Holland's *No. 6* became father to the famous "A" boats prior to World War I.

# Diving for Science

The other significant development in the undersea boat story during the latter nineteenth century that carried over into the present was started by the American, Simon Lake. As a boy, Lake was intrigued by the romantic picture created in Verne's *Twenty Thousand Leagues Under the Sea,* to the point where he began to design a series of improvements on the fictitious *Nautilus* of Captain Nemo.

He chose to become an inventor and left school before the age of seventeen. He built his first submarine in 1894, a scaled-down version of a larger one already conceived. *Argonaut Jr.* was a 14-foot long vehicle constructed of pitch pine and canvas, with several unique features. The most intriguing of these for the passengers was a bottom hatch that could be opened for viewing, sampling, or diver entry and exit. Equally interesting was the presence of large wooden wheels, two in front and a smaller rear one, which were connected by a bicycle chain to an interior crank mechanism. In this way, the *Argonaut* could crawl along the bottom. After several successful trials, Lake demonstrated his novel boat to a crowd of onlookers including the mayor and officials of his home town, Atlantic Highlands, New Jersey. He managed to raise both interest and money for his submarine ideas. Lake's real goal was to woo the U.S. Navy and sell his advanced ideas to the government, so his next move was to build the full-sized *Argonaut* in the same Baltimore shipyard where the Holland *Plunger* was being constructed for the U.S. Navy. *Argonaut I,* a 36-foot iron vessel with 7-foot cast-iron wheels, was launched in Baltimore in 1897. It was propelled by a 30-HP gasoline engine that breathed through a long hose and a float—an early snorkel.

This larger version also had the swim-out lock which completely fascinated everyone who made a dive. Lake was a born promoter and cleverly arranged for the press to join him on a dive. Over twenty reporters went on one dive off New York and watched through the bottom lock as a suited diver left *Argonaut,* gathered some shells, and returned to the submarine. The newspaper coverage of the Lake submarine was outstanding, but not one of the U.S. Navy officials or Washington congressmen was the least bit interested. Lake could not even get a chance to demonstrate his vehicle to them. Perhaps it was that the *Argonaut* was too unconventional for a Navy barely coming around to the Hol-

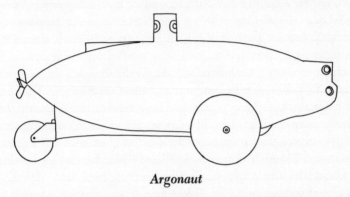

*Argonaut*

land submarine. Lake also strongly favored the level dive, rather than pitching bow down and plunging, and the Navy could not accept this approach. Further, the idea of a craft crawling on the bottom made naval officers uneasy. They pictured the submarine plunging over a cliff, becoming too heavy to surface, and being crushed by pressure.

Lake's ideas were gradually accepted, but much too slowly for his liking. He continued to show what his submarines could do. In 1898, he slipped quietly into Hampton Roads, Virginia, where the U.S. Navy had placed mines to discourage any attack by the Spanish. He inspected them and showed how he could even stick them onto ships. The Navy officials were not amused or interested. Following this, he made an unheard of voyage of over 1,000 miles from Norfolk to New York. Much of it was submerged. He proved the seaworthiness of his boat during this venture.

Simon Lake raised some investor capital from Bridgeport, Connecticut, and in 1902 built a 65-foot submarine, the *Protector*. With it, he made the first dives under and through ice in Newport Harbor. Still the Navy paid no attention. The Lake Torpedo Boat Company was in difficult times. Electric Boat, Holland's company, was suing for patent infringement and Lake had few funds left. Although he was not interested in military submarines, he found as his only customers the belligerents, Russia and Japan. He managed to sell his *Protector* to the Russians for $250,000 but as part of the deal he had to live in Russia for seven years.

Upon returning to America, Lake again took up the competition with the Holland Company and their successful submarines built for the U.S. Navy. Lake was now forty-three years old, but hardly slowed by his lack of success. He was a man of medium height, stocky, and recognized by his fiery red hair and brilliant blue eyes. Through diligence and hard work, he and his workers managed to win a Navy contract, and in 1912 they delivered their first submarine to the U.S. Navy. It was a great triumph for him. The *Seal* was a 161-foot submarine with a 30-foot beam. She was the largest and fastest vessel in the sub fleet, making ten knots submerged. *Seal* became the SS-19½, the only fractional number ever assigned a submarine. The Lake Torpedo Boat Company built other submarines for the Navy toward the close of World War I. Yet Lake himself was not interested in the submarine as a weapon and he continued to develop submarines for scientific and explorative endeavors. He devised plans and hardware for treasure salvage, but was never able to undertake the salvage of ships such as the *Lusitania*.

One of his most interesting vehicles was built for exploration of the bottom. In 1932, he launched the *Explorer*, which was a perfection of all his early small vehicles. *Explorer* was a 10-ton, 22-foot long vessel with electrical power supplied by cable from a surface ship. The two crew members had separate ports in the conning tower, and there was room for two divers. Lake

# The Founding Fathers

was fascinated with the idea of gathering oysters, salvaging cargo such as ore and coal, and carrying out underwater engineering projects. Unfortunately, little appears to have been done with *Explorer* although its design resembles more modern vehicles.

Simon Lake was convinced that submarine cargo vessels were practical. Along with Fridtjof Nansen, the Norwegian explorer-oceanographer, he saw that the polar route under the ice could be a practical one. He designed several submarine cargo vessels as early as 1898, but at the turn of the century found no one even slightly interested. His cargo idea was stolen by the Germans during World War I when they built the *Deutschland* and the *Bremen.* The *Deutschland,* the world's only successful submarine cargo ship, made two trips to the United States in 1916. Lake was unable to collect anything from the Germans for his ideas.

Lake, working actively on his inventions at the age of seventy-four, was more intent on the invention and the success of his projects than on himself. But few Americans knew of him, and in 1945, at the age of seventy-eight, he died a poor man. His contributions to submersible development were great, however. He had made the submarine steer and dive stably, designed its periscope, proved a lock-out hatch worthwhile, and generally established a submarine hull configuration. His ideas have helped the subsequent development of the research submersible as much as those of any one man.

One of the first purely scientific adventures undertaken by submarines, one that helped show the way for the research submersible, was the Wilkins Expedition in 1931. Sir Hubert Wilkins, an Australian explorer of the polar regions, had decided it was possible to reach the North Pole under the ice by submarine. He was convinced that significant scientific contributions could be made by this voyage. Such measurements as the bathymetry of the Arctic Ocean, magnetic and gravity observations, and the collection of water samples from various depths were among the

41

goals of the expedition. Wilkins believed it might be possible to set up weather observation stations near the pole.

It is no coincidence that the project engineer for the submarine turned out to be Simon Lake. Unfortunately, the only ship available was an aged O-class submarine built in 1917 by The Lake Submarine Company and about to be scrapped under a World War I treaty. Lake and a partner, Sloan Danenhower, agreed to modify and prepare the *O-12* for Wilkins. Danenhower was elected to skipper it. Wilkins was then a world famous explorer and, with the cooperation of Woods Hole Oceanographic Institution and the Norwegian Geophysical Institute, had no difficulty enlisting oceanographic scientists. Dr. Harald Sverdrup of Norway was chosen Chief Scientist. Later Sverdrup, as Director of Scripps Institution of Oceanography, helped to place American oceanography in a position of world prominence.

The submarine was taken out of moth balls and converted for Arctic service. Christened *Nautilus* in honor of Verne's *Nautilus*, she was 175 feet long with an 18-foot beam. As a military submarine, she would be considered small by today's standards, but for scientific exploration, she was the largest vessel ever put to that use. Her surface speed was 14 knots with a range of 2,930 nautical miles on her two 500-HP diesel engines.

Lake incorporated a number of his inventions on the submarine. One of these was a major modification for the under-ice mission. An ice drill was built which permitted drilling through thin ice patches to allow air in for diesel operation and daily battery charging. The drill tower measured 30 inches in diameter and was over 15 feet long. The entire assembly rotated through an elaborate tooth and gear assembly. One of the advantages was that the tower was large enough for a man to crawl through to the surface. Lake also installed a lock-out chamber similar to the one in the *Argonaut*. Through this entry, scientists lowered water bottles, corers, and plankton nets while the submarine was submerged.

After many delays, the *Nautilus* left Provincetown, Massachusetts bound for England. When she was several days out, in the midst of a storm, engine trouble developed. Captain Danenhower radioed for assistance, and an American battleship, U.S.S. *Wyoming*, came to the rescue and towed *Nautilus* and crew to England. Other mechanical problems plagued the ship until finally, in the ice pack north of Spitzbergen, her diving planes were accidentally carried away. Even so, Wilkins insisted that the expedition carry out all possible observations. One shallow dive beneath the ice was ventured, although sea conditions were not ideal. Wilkins felt the expedition was a failure, yet a number of worthwhile soundings, several plankton tows, and core samples were taken, and the first gravity readings were made in the Artic. The ship came within 500 miles of the North Pole.

Reflecting on the ill-fated voyage years later, Wilkins stated, "We had collected sufficient experience and knowledge to enable someone, someday, to complete a journey under the Arctic ice." In the last year of his life, Wilkins saw his ultimate goal accomplished by the *SSN 571* nuclear submarine, also named the *Nautilus*, when she traveled beneath the North Pole in 1958.

The next important link in the submersible chain is the scientific contributions and diving technology developed by Professor Charles William Beebe and Engineer Otis Barton. Beebe was known in his earlier years as a biologist and more specially as an ornithologist and was on the staff of the New York Zoological Society beginning in 1899. William Beebe, as he later became known to millions, had spent much time collecting deep sea creatures off Bermuda in the 1920s as an extension of his naturalist interests. Moreover, he had also been one of those few to experience the thrill of seeing the brilliant colors of the shallow underwater reefs in many clear water areas. He began to realize how very limited man was in his ability to venture to any depth in the ocean, but he knew that investigations could be done at depths greater than the 300 feet that man had reached in limited diving suits.

Beebe recalled discussions he had with President Theodore Roosevelt on the subject, in which he described a pressure cylinder for a manned chamber while Mr. Roosevelt preferred a sphere. Later in 1927 and 1928, Beebe worked with various plans for a "deep-sea cylinder" but when he found that Otis Barton had already built a steel sphere he turned gladly to this form. The spherical chamber that was to become the bathysphere had been designed by Captain J. H. Butler and financed and constructed by Otis Barton.

Beebe commented on his faith in the deep diving bathysphere that was to take him a half mile into the mother ocean, "Never for a moment did any of us admit the possibility of failure . . . my hopes of seeing a new world of life left no opportunity for worry about defects." Barton's sphere, designed by the team of Butler and Barret of the Cox and Stevens Company, was 4 feet 9 inches in diameter and was made of a single casting of 1¼-inch thickness steel with a total weight of 5,400 lbs. There were three ports of fused quartz—the best optical material known for the passage of all wavelengths of light. The ports were 8 inches in diameter and 3 inches thick. Only two of the quartz windows

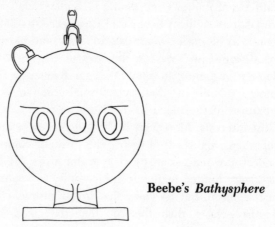

Beebe's *Bathysphere*

were used, one for the main head-lamp for underwater illumination and the other for observation; the third aperture was plugged after the window was cracked during installation. Two other windows had previously been broken during assembly and tests. Opposite from the ports was the hatch, or as Beebe said, "the entrance, politely termed the door." The hatch turned out to be a massive 400-pound steel plate that was bolted into place with ten large nuts before each dive. The passage was a 14-inch opening that allowed the divers to gingerly slide through. The inside atmosphere was serviced by a crude life support system in the early series of dives. Later the hand fan for air circulation was replaced for the deep dive by an electric fan.

There is little doubt that the bathysphere represents the first serious deep submergence chamber. The sphere design, size, ports, life support system with oxygen supply, all are the clear ancestors of equipment that made deeper scientific dives in the following two to three decades.

Unlike the research submersibles that were to follow, Beebe's depended on a tethered system that used a cable, since the principles for a free, deep vehicle had not emerged. Thus, a cable weighing two tons, 3,500 feet long of nontwisted type steel, with a ⅞-inch diameter, was used to suspend the sphere in midwater and let two men peer out at what human eyes had never seen in the three billion years of the ocean's existence. The cable greatly aided in letting Beebe make valuable scientific observations, since it meant that electricity for lighting could be supplied along a separate cable married to the strain member, and that telephone communication was possible for transmitting all biological observations to an assistant who transcribed all these reports.

Beebe had an ideal type of training and experience that led logically to the first deep diving effort. His thirty years of biological work with two more years concentrated in the ocean near Bermuda gave him a thorough grounding in understanding the underwater communities and animal forms normally obtained by dangling nets and dredges from the surface into the water column. He knew how to identify the animals of the deep and he was to find that through this new "window" he would see animals in abundance never imaginable from the sparse evidence of net hauls.

The bathysphere made thirty-five dives altogether in its 1930, 1932, and 1934 seasons. Most of these were observational in nature in which Beebe described for all the world what it was like to descend in the 4-foot 9-inch ball into a completely new realm. More importantly, he added valuable information to the understanding of the animal population of the mid-depths, the depth of light penetration, and the measurement of types of wavelength extinction. In the first dive series, Beebe and Barton achieved a quarter-mile depth (1,426 feet) on dive number seven. To test the sphere at this depth they had lowered it to 2,000 feet unmanned. On several earlier shallow dives the hatch had leaked slightly, so they decided to seal it with white lead before screwing the ten bolts. Further disconcerting events that might have upset these fathers of underwater exploration or stopped them from diving were the several electrical short circuits, and the "hosing" or intrusion of the rubber cable through the stuffing box into the sphere. However, unruffled, they made repairs, corrected difficulties, and finally made dive seven successfully.

William Beebe captured impressions and emotions that helped many people to live this rare first experience venturing to the edge of the abyss:

> At the very deepest point we reached I deliberately took stock of the interior of the bathysphere; I was curled up in a ball on the cold, damp steel, Barton's voice relayed my observations, our hand fan swished back and forth through the air . . . I sat crouched with mouth and nose wrapped in a handkerchief and my forehead pressed close to the cold glass . . . there came to me at that instant a tremendous wave of emotion, a real appreciation of what was momentarily almost superhuman, cosmic, of the whole situation;

our barge slowly rolling high overhead on the blazing sunlight, like the nearest ship in the midst of the ocean, the long cobweb of cable leading down through the spectrum to our lonely sphere, where, sealed tight, two conscious human beings sat and peered into the abyssal darkness as we dangled in mid-water, isolated as a lost planet in outermost space. Here I was privileged to peer out and actually see creatures which had evolved in the blackness of the blue midnight which, since the ocean was born, had known no following day.

Beebe recalls the value of these early dives by commenting, "As fish after fish swam into my restricted line of vision—fish, which, heretofore, I had seen only dead in my nets—as I saw their colors and their absence of colors, their activities and modes of swimming and clear evidence of their sociability or solitary habits, I felt that all the trouble and cost and risk were repaid manyfold." He felt that being at the spot from which he had hauled many nets enabled him to appreciate future net collection. He could see convincing proof of the abundance of organisms—and that a great many of them avoided the nets or were not represented in the nets. Further in the transition zone from the shallow nearshore down the steep rugged slope to deep water, it was difficult or impossible to trawl nets which had a small chance of capturing fish.

The bathysphere dives were extremely valuable in contributing to the evolution of the deep submersible. They were not without some relatively narrow escapes by Beebe and Barton who appeared somewhat nonchalant about the dangers involved. In 1932, on the second series of dives, the tug and barge that staged the bathysphere were brought close to shore, and then they began diving down the slope beginning at ten fathoms. This sort of "contour diving" in which the bathysphere was kept several fathoms off the bottom by quick orders over the phone to the workmen at the steam winch must have been exciting. On one occasion Beebe spotted an enormous coral crag fifty feet above them in the path of drift. He shouted the order to raise the ball immediately. The winch responded and the bathysphere with its two men shot upwards grazing the coral. Any hesitation on the part of the winchman and the ball would have been plucked off the wire. This sort of near miss did not deter these heroic divers in the least. Concern about the cable and the chance of its breaking is certainly one of the chief disadvantages of the bathysphere especially as greater depths are attempted and the cable weight increases. Clearly the great depths of 10,000 to 20,000 feet could not be explored by a tethered or pressure cabin.

In 1932 the season wound up with a dive to 2,200 feet with more exciting observations of the mid-water fish, but it was not without several harrowing experiences which the pilot and crew of the bathysphere took in their stride. Professor Beebe decided to place a new spare quartz port in the blank third aperture. To test the installation the bathysphere was lowered on an unmanned test dive to 3,000 feet. When it returned he saw at once something was very wrong. The sphere was almost full of water and contained only a little air at very high pressure—a thin needle of water and air shot across the face of the port. The deck was cleared of people and Beebe carefully loosened the center wing nut of the hatch while the high pressure leak whined and shot steam. Beebe comments in his book, "Suddenly, without the slightest warning, the nut was torn from our hands, and the mass of heavy water shot across the deck like a shell from a gun. The nut lodged in a steel winch frame leaving a half-inch dent in the harder metal, while a solid cylinder of water roared across the deck." Undaunted the crew repaired the leak and replaced the metal port and readied the bathysphere for use.

The 2,200-foot dive that followed was broadcast to the world by NBC. Professor Beebe and his diving bell had quickly become known to Americans interested in his adventures. NBC featured a live broadcast in which his telephone observations were relayed over 3,000 feet of cable to the radio audience.

Dive number thirty-two off Bermuda, was the culmination of

the bathysphere's record when Beebe and Barton reached 3,028 feet on August 15, 1934, setting a manned diving record that stood for over fifteen years. The observations Beebe made and his subsequently published book, *Half-Mile Down,* are significant contributions to both science and technology and are the first manned reports on the deep ocean.

It was an exciting dive. At the deepest point a scant fifteen turns of cable was all that remained on the winch drum, which was half-naked to the core, and the ship's captain, fearing the strain on the end of the cable, refused to let another foot out. The 2½-ton sphere and 2 tons of cable dangled precariously under a total pressure of 1,360 pounds to each square inch—half a ton at half a mile.

Beebe speaks of three outstanding moments in the mind of a bathysphere diver as being the first flash of animal light, the level of eternal darkness, and the discovery and description of a new species of fish. He relates in the log:

> At 2,450 feet a very large dim, but not indistinct outline came into view for a fraction of a second, and at 2,500 feet a delicately illuminated ctenophore jelly throbbed past. Without warning, the large fish returned and this time I saw its shadow-like contour as it passed through the farthest end of the beam. Twenty feet is the least possible estimate I can give to its full length. . . . I could not see an eye or a fin. For the majority of the "size conscious" human race this would, I suppose, be the supreme sight of the expedition.

Beebe thought it could be a cetacean or possibly a whale shark. Submersible pilots have since seen similar large animals at deeper depths that have been termed "sea monsters."

Professor Beebe, in his beautifully composed style that so captured the reader and informed his fellow scientists, summarized his diving experiences prophetically in 1934:

> Whenever I sink below the last rays of light, similes pour in upon me. Throughout all this account I have consciously rejected the

scores of "as ifs" which sprang to mind. The stranger the situation the more does it seem imperative to use comparisons. The eternal one, the one most worthy and which will not pass from mind, the only other place comparable to these marvelous nether regions, must surely be naked space itself, out far beyond atmosphere, between the stars, where sunlight has no grip upon the dust and rubbish of planetary air, where the blackness of space, the shining planets, comets, suns, and stars must really be closely akin to the world of life as it appears to the eyes of an awed human being, in the open ocean, one half mile down.

While Professor Beebe was making his first bathysphere dives on the end of a long string, another professor was hard at work on the final equipment that would allow him to ascend into the stratosphere to explore the nature of cosmic rays above the limitations of the earth's atmosphere. Auguste Piccard, one of the great contributors to twentieth-century science and technology, was first a recognized balloonist and explorer of the upper atmosphere. His solid background in physics and technology aided in his inventions that made great areas of the world accessible to man.

Professor Piccard in 1931 and 1932 made balloon ascents to 53,000 and 55,000 feet to measure characteristics of the sun's radiation unaffected by the heavy screening qualities of the earth's atmosphere. To do this he had designed and had built in Switzerland an aluminum gondola with two hatches. The gondola was a pressure cabin that would allow the observers to ascend to an altitude where the pressure is one-tenth that of sea level—2.992 inches of mercury or 1.47 pounds per square inch. The balloon ascents were financed by a foundation in Belgium—Fonds National de la Recherche Scientifique. The balloon was baptized the *FNRS* and so it was that Piccard made his first exploration into the stratosphere in *FNRS-1* to a calculated height of 53,000 feet or 10 miles above the sea. The ascent was fraught with serious difficulties and had it not been for Piccard's ingenuity, he and his companion might have perished. After this success, Piccard went on to make further discoveries of the physics of the upper atmosphere and stratosphere.

Once one understands the principle of a pressure cabin and supporting balloon, it's quite simple to see the analogy to the bathyscaphe (Greek for deep ship). And since the valuable experience gained by Beebe and Barton was so closely related to Piccard's, few people question his decision to leave ballooning to experiment with underwater craft. He makes an interesting statement in his book, *Earth, Sky, and Sea,* that many people may have overlooked in following the evolution of the research submersible. First, he explained that he developed the concept of the bathyscaphe when he was a student at Zurich Polytechnic School about 1920. He had read in the account of the expedition of the oceanographic research vessel *Valdivia* about what happened to the netted abyssal fish. Unable to survive at the much lower pressure and higher temperature, they expired, and their little-understood characteristic phosphorescing organs extinguished. Piccard realized then that scientists would have to go to the great depths of the ocean to see these animals alive. He reflected on that moment:

> It must be possible, I said to myself, to build a watertight cabin, resisting submarine pressure and furnished with port holes to allow an observer to observe a new world. This cabin would be heavier than the water displaced. It would be necessary in complete analogy of the free balloon, to suspend it from a large vessel with a substance lighter than water . . . the fundamental principle of the bathyscaphe was born.

It was later in his career when Piccard desired to investigate the cosmic rays in the upper atmosphere that he returned to the idea of the submarine pressure cabin for solution to the problem of the low pressure at high altitudes. He said, "The evolution of my thought is clear. Far from having come to the idea of a submarine device by transforming the idea of the stratospheric balloon, as everyone thinks, it was on the contrary, my original

51

conception of a bathyscaphe which gave me the method of exploring the high altitudes. In short, it was the submarine which led me to the stratosphere."

Piccard began the design of the bathyscaphe by seeking support, as he had for *FNRS-1*, from the Belgian Fonds National. He turned to the design of high pressure portholes and electrical fittings that could withstand 1,600 atmospheres, or equal to a depth of 10 miles. Before the experimental work could proceed far, World War II closed in on Europe and all such effort was curtailed until 1945.

At that time Piccard continued with his plan to build an "undersea balloon" and, with renewed financial aid from the same Belgian agency, he and his Swiss group produced the world's first deep submersible in 1948. He named this strange looking device *FNRS-2* in memory of his first stratospheric balloon. Economics of the project forced him to design the submersible to be carried aboard a ship and not be towable. This fact later contributed to the premature damage to the vehicle.

The *FNRS-2* followed the balloon principle with a sizable hollow float, oval-shaped, measuring 22 feet by 10 feet, and with a volume of 1,059 cubic feet. Six tanks inside the float contained 6,600 gallons of gasoline which was lighter than water and thus had sufficient buoyancy to support the pressure cabin just as a gas bag does a gondola.

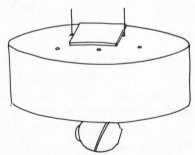

**Piccard Bathyscaphe *FNRS-2***

The most important component and the most difficult to engineer was the cabin or pressure sphere. It had to be large enough to house two men and scientific equipment and yet be of minimum size to limit the weight. The cabin is the principal source of weight in a submersible, since it must have walls thick enough to withstand with safety the pressures at depth. While Beebe chose a 4½-foot-diameter cabin and found it extremely uncomfortable, Piccard decided the best size-to-weight compromise was at 6⅗-foot diameter or 2 meters. The resulting sphere would weigh 5½ tons. He calculated a slight increase to 6 feet 10 inches would add 3,520 pounds to the total, thus forcing the float to increase considerably to hold more gasoline for buoyancy. The crew would have to fold themselves into a 6-foot 7-inch sphere to save on weight.

The design of the steel sphere was complicated by the penetrations for portholes, wire passages, and a hatch. The optimum thickness was 3.54 inches, increasing to 5.9 inches thick around the penetrations to allow for increased stresses. Professor Piccard used a design depth of 10 miles so that at the proposed operating depth of 2½ miles, there was a safety factor of 4. The sphere was machined out of two hemispheres joined by an equatorial joint.

The key to the successful design of the submarine which would serve the observing scientist was a port or window that afforded wide-angle viewing at the depth of 10,000 feet. These windows had to withstand forces of 500 tons. No theory for such a design existed. The port used by Beebe and Barton was of fused quartz which, as they had seen, was unpredictable and subject to cracking. The solution that Piccard saw in 1939, and the resulting design, was a significant breakthrough for submersible use that has continued virtually unchanged to the present. Piccard and a colleague, Professor Guiltisen, experimented with plexiglas, then quite new, and designed a cone-shaped port 3.9 inches inside by 15.7 inches outside and 5.9 inches thick. In high pressure testing,

Diving for Science

they saw that when the plastic was overloaded beyond its limit of elasticity, it merely passed the excess stress to adjacent parts with equal distribution. This cone-shaped port of plexiglas has since been used for all deep submersibles with complete success. Piccard commented in *Earth, Sea, and Sky,* "These windows are perhaps the finest feature of the bathyscaphe."

The bathyscaphe *FNRS-2* used iron shot for ballast, as well as several tubs of scrap iron held by electro-magnets, gravel in tanks, and finally two heavy storage batteries which could be electrically released. The ballast was vital to enable the bathyscaphe to descend and then, at any time, to drop shot or, in an emergency, other ballast for ascent to the surface. Among the other important features of *FNRS-2* was the electrical source for lighting and propulsion. The batteries were lead acid type, oil-filled, and compensated for sea pressure. These two batteries supplied 900 ampere hours. As with the port design this type of pressure-compensated battery case has been used successfully for most vehicles since 1948.

The *FNRS-2* was shipped aboard a 3,500-ton cargo ship, the *Scaldis,* and made ready for the trip to Dakar. They left on September 15, 1948. Dakar was the closest place where the weather was generally fair and where the water reached 3¾ miles in depth. The actual deep spot, however, was about a day's steaming from Dakar. By mid-October, the Piccard diving crew was joined by Captains Cousteau and Taillez of the French Navy, with the oceanographic ship *Elie-Monnier.* On October 26, the *FNRS-2* was lifted from the hold, gently placed in the water, and then the empty float was filled with 7,040 gallons of gasoline. While this rather delicate operation was going on, the occupants were already inside. Calm seas were essential. After much waiting and a failure of the telephone line, the *FNRS-2* slowly departed the surface and sank to the shallow bottom of 14 fathoms which they had chosen for the first check-out dive. This first dive was made by Professor Piccard and Dr. Monod, who spent

54

twelve hours shut up in the small sphere. They were well supplied with air freshened by the Drager system which was exactly like that which Piccard used in his balloon.

The next dive was to be an unmanned one to test a depth of 825 fathoms, using an automatic pilot device that would deballast the vehicle either after a certain time, or when it touched bottom. A spot of about 800 fathoms was selected, but due to delays the *Scaldis* drifted into shallow water before the *FNRS-2* could be launched. They towed the vehicle awkwardly into deeper water. Finally, with little over an hour left on the timer released clock, Piccard let go of his precious invention, wondering if the automatic system would work and return his child to the surface. They all waited with anxiety. Then, suddenly, in the short time of twenty-nine minutes, it was back on the surface. Piccard could not believe the round trip of 1,500 fathoms could have been made so quickly. While they prepared to pump out the gasoline and bring her aboard, the wind and sea came up and night fell suddenly. The waves washed over the nearly submerged bathyscaphe making it impossible to attach the pump-out hose. Fearing for the safety of the divers working around the boat, Cousteau, the safety officer, decided to secure operations. *FNRS-2* could now only be towed behind *Scaldis*. But *FNRS-2* was hardly designed for towing and she balked, pitched, and yawed. Even at a snail-slow pace, the waves beat and bent her float. They stopped and decided to jettison the gasoline into the sea, hoping to be able then to hook her and lift her out of the water, but still it was too rough. So, for the rest of the night, the valiant submersible thrashed along behind *Scaldis* in the mountainous sea, and was pounded and damaged beyond repair.

At dawn, in the lee of an island where it was calm, Auguste Piccard climbed excitedly into the cabin of the vehicle after it had been brought aboard. To his great joy, he found that the recording depth gauge showed 759 fathoms or 4,554 feet. The bathyscaphe actually had been to the bottom, making a descent speed of 5 feet per second and a fantastic ascent of 6 feet per second.

While Professor Beebe still retained the record from 1934 for a manned dive of 508 fathoms or 3,048 feet, the *FNRS-2* had proved that much deeper depths were attainable. Interestingly, Otis Barton, designer of the bathysphere, had built an improved sphere, the *Benthoscope*, in 1948. In October 1948 he dived the 7,000-pound tethered sphere to 4,488 feet off Santa Cruz, California. This dive held the world manned record for several years before bathyscaphes began their push to the oceans' greatest depths.

Although this was the only dive of *FNRS-2* and a disappointing one, 1948 marks the beginning of an era of true deep submersibles which for the first time were free from the surface and able to dive to significant depths. With a few minor modifications, the subsequent vehicles of Professor Piccard and his son, Jacques, constitute important ancestors of more modern vehicles. *FNRS-2* was the basic prototype for subsequent bathyscaphes *FNRS-3, Archimede, Trieste I, and Trieste II.*

The final vehicle that belongs in any of the historical accounts of progenitors of submersibles is the *Trieste*. Its lineage traces directly from the *FNRS-2* and it probably is the best-known research submersible because of its ten-year history of deep dives culminating in the world's deepest dive, which was made in the Pacific Ocean.

The lack of funds for *FNRS-2* had prevented Professor Piccard from designing it to be towed, so he returned to his first design of 1938 in which the bathyscaphe had a cylindrical shape. Armed with these plans, in 1949 he trudged off to see his friends at the Belgian Fonds National. To his dismay, he found that the public was disinclined to support further submersible building. Instead, a collaborative arrangement was arrived at between Fonds National and the French Navy to build a working deep submersible —dubbed *FNRS-3*. It was a cumbersome arrangement of groups in which Professor Piccard did not function well since he did not have final authority. The *FNRS-2* hull had been preserved and was used as the hull for the new boat, and by 1952, progress was slow and frustrating. Fortunately for Auguste Piccard and son Jacques, who was then helping him, a proposal was made by the city of Trieste to construct a new bathyscaphe there. This arrangement was also a collaboration—but in this undertaking Piccard was solely in charge. The city of Trieste provided the industrial know-how and machinery, while Piccard's home country of Switzerland put up much of the money.

As *Trieste*, as the the new vessel was named, began to take shape, it was clear that she was closely related to *FNRS-2*, but with marked improvements that would permit easier towing at sea, and also crew entry and exit in rough seas.

The most obvious change was in the float which was cylindrical and measured 49 feet in length, twice that of *FNRS-2*. The float had twelve compartments for gasoline and two tanks, one at each end, for air to give buoyancy while on the surface. Between each compartment was a corrugated bulkhead for strength. A vent system was installed to allow air from the end compartments to escape when seawater was flooded in for submergence. Directly or indirectly all compartments of the float were in communication with the sea. As with the free balloon, the bathyscaphe needed a control valve to release gasoline to check the rate of ascent or descent, and Piccard developed an ingenious electrically-controlled valve that was activated by the pilot.

The reverse process from venting gasoline, that of unballasting, was improved from the *FNRS-2* which had tons of scrap iron ballast held by magnets. *Trieste* had only one system of deballasting—which used nine tons of iron shot pellets. The shot were loaded into two metal tubs, together weighing 2 tons. At the base of each tub was a magnetic valve for releasing shot. Each of the tubs was suspended from an electro-magnet. In an emergency the tubs with shot inside could be released by turning off the electric energy that held them magnetically to the float.

Since the bathyscaphe was to be left in the water, and the crew would have to enter and exit in various sea states, Piccard designed a tower and entry tube from the surface to the cabin. This tower was 25 inches in diameter, and had a ladder of about 13 feet leading to the hatch of the pressure cabin. Once the crew was in the cabin and the lower hatch was secured, the tower chamber was flooded. Under normal conditions, the tender vessel could pump the water out when the bathyscaphe returned to the surface, but a set of compressed air bottles was placed aboard *Trieste* to allow the crew to expel water from the entry chamber. The hatch of the cabin had a plexiglas port so that the observer could look into the antechamber, which had a large plexiglas window (24″ x 34″) for viewing aft, and see if the shot was being dropped.

*Trieste*'s pressure sphere, in two parts, was forged from a casting of greater malleability than the *FNRS-2* sphere, making it easier to work with. The initial forgings of each hemisphere weighed 10.8 tons. They were then put on a lathe to machine them accurately and in this process 5 tons were shaved off from each. The inside dimension of the finished cabin was 6½ feet in diameter, the same as *FNRS-2*. The wall thickness was 3½ inches, with a single porthole measuring 15 inches outside and 4 inches inside. The hatch diameter was 16.9 inches, with a second port in the hatch door. The principal port looked down at about 20° from the horizontal. As with the earlier hull, the two hemispheres were joined by a circumferential ring that clamped the surfaces together without a gasket. Piccard realized that a gasket between the hemispheres would flatten and extrude under high pressure and thus only a precisely machined surface without a gasket would suffice. Finally, to solve the most difficult problem for a deep submersible—the passage of cables from outside through the hull—Piccard designed twelve penetrators radially spaced around the main port. These had to be completely water-tight since at a depth of 15,000 feet water from the outside could flow in at 26 gallons per second, giving occupants only seventy seconds of life at that pressure. A copper tube filled with densely packed dehydrated asbestos was used. Small conductors were insulated within this asbestos. This type of penetrator, known as a "Pyrotenax" cable, allowed Piccard to eliminate the oil-filled contactors for the switching on of lights since he was able to bring in larger-gauge wires. The oil-filled contactor boxes had caused problems on *FNRS-2* and *FNRS-3* and still are subject to failure in more modern vehicles. Besides electrical conductors, *Trieste* also had air ducts and tubes for pressure gauges.

For life support within the cabin, Piccard turned again to the same group that had built a system for *FNRS-2*. Dragerwerke, a German company, designed a closed-circuit oxygen system that supplied sufficient regenerated and oxygen-enriched atmosphere for three men for at least twenty-four hours. For emergency periods on the surface, if the crew for some reason had to remain in the bathyscaphe for several days, a two-tube snorkel to the surface was designed, which used an electric ventilator.

The electrical supply for *Trieste* came from a silver zinc battery of 900 amp-hours capacity that was about one-quarter the size and weight of the lead acid battery used in *FNRS-2*. An added advantage was that the silver zinc battery could be placed within the cabin, thus avoiding pressure compensation with oil. It also could be charged in place without giving off explosive gases.

One of the large demands on the electrical supply was the external lights so important to underwater viewing. A specially designed projector lamp with an incandescent 1000-watt bulb was devised to handle this. These relatively high-powered lamps were placed forward of the view port to sidelight the bottom objects and avoid the scattering effects caused by particles in the water.

Other instruments aboard were four depth gauges, a vertical speed indicator, radio, hard-wire telephone for surface talking, and internal lighting of low level to allow instrument reading in the dark.

Then came the first dives. The "getting wet" process, which is always a frustrating experience of adjustment and modification, was to be carried out at Castellammare di Stabia in the Gulf of Naples. The float was transported by truck in January, 1953 at a crawling speed of 9.3 mph, taking 11 days from Trieste around the Adriatic and Mediterranean shoreline. At the same time, the pressure sphere made its way also by truck from Terni. Professor Piccard was held in such high regard that neither trucking firm thought of charging for these laborious journeys. In the name of science, it was an honor for them to contribute their services.

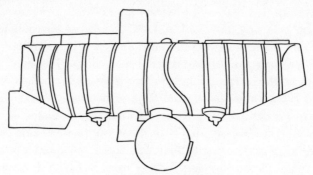

Bathyscaphe *Trieste*

By August of that year, the *Trieste* ran up her colors and was blessed instead of having the traditional champagne ceremony as she was quietly immersed without gasoline into the sea. Later, Esso put aboard 18,920 gallons of a special grade of gasoline.

Piccard made the first dive with great expectation, although it was in shallow water, only 132 feet deep, in case some need should require divers. Jacques Piccard stayed on the surface, communicating by hard-wire telephone. The air tanks were flooded and slowly the *Trieste* sank into the clear blue water.

About 40 feet from the bottom, she stopped. There was a cold, more dense layer of water normally found in such situations off shore. Instead of waiting for the gasoline to cool, the elder Piccard decided to return to the surface for more ballast. This done, he descended to an unexciting mud bottom without much life. After maneuvering and checking controls, he surfaced. The dive was hardly a record or a scientific accomplishment. Piccard pointed out the first dives of a new vehicle seem to the casual bystander a bit silly, when expensive craft designed to dive several miles are puttering about in water less than a hundred feet deep. But these training and learning dives pay off later when difficult situations in the frequently rough open water allow no time for learning from mistakes. The return for more ballast could not always be done so easily.

After several shallow check-out dives in which almost no modifications were necessary, the Piccards decided it was time to make a real ocean dive. Professor Piccard was so confident of his design for the through-hull connections and view ports, the most obvious points of possible leaks, that he felt it unnecessary to make an unmanned descent to a test depth. *Trieste* may be the only deep submersible that went to deep depths without first performing a pressure test of the unmanned hull—it is a tribute to Professor Piccard's exacting design and fabrication.

A site was chosen south of Capri with a depression of 600 fathoms (3,600 feet). After some ballasting problems where shot was dumped by a faulty cable, Piccard decided to dive on August 25, 1953, with only one shot tub operating. To correct the difficulty would have meant a delay of a week or more for towing back to port and removal of gasoline from the float and the float from the water. We can see from this that as remarkable a boat as *Trieste* was in her ingenious design, she was not at all favorable from the logistics standpoint and required a tremendous support effort.

The dive began and Piccard related, as had Beebe so beautifully described twenty years earlier, that the light outside the porthole slowly faded into a brilliant blue purple, and then a finally imperceptible violet black. At 250 fathoms, they switched on their lights to see a few flittering small fish. As expected the gasoline continued to cool in the ever-colder surrounding water, making the bathyscaphe heavier, and it picked up speed. At this rate of descent, it became impossible to observe any of the marine life through the porthole. Although the divers were fascinated by the few fish they saw, their mission on this dive was to get to 550 fathoms and prove the bathyscaphe a sound vessel; so they resisted the temptation to dump shot and slow their descent rate until they came close to the bottom. But they miscalculated their speed (the vertical speed indicator had not been installed) and they landed too fast; without knowing it they had buried the cabin in the soft ooze of the bottom. Professor Piccard had long awaited this triumphant moment to look at living organisms on the bottom which no human had ever seen. Now the port was thoroughly blocked, as the cabin rested 4 feet in the grey blue clay. What should he do? The only choice was to drop shot and return. The entire load of the aft tub shot was released. They waited while the shot dropped at 110 pounds per minute. Finally, with a lurch, the *Trieste* pitched forward. Professor Piccard lunged to the port to see if he could see anything with the 5,000 watts of light, but the cloud of silt stirred by the sphere pulling out of the mud was too murky. They ascended slowly at first, then picked up speed as the gasoline expanded. After about forty-five minutes, the *Trieste* was on the surface in the bright sunlight with her first deep dive successfully accomplished even if the chance for bottom viewing was spoiled. Although again, no scientific observations were made, this experience was groundwork for later significant discoveries by the first bathyscaphe. Piccard was ready to attempt a much deeper dive —the first step toward diving to the deepest spot in the world ocean.

The dive site chosen was off the Island of Pouza, about 60 miles from the base of Castellammare where there appeared on the charts a broad sandy plain undulating from 1,650 to 1,760 fathoms. Piccard wrote of this area, "It was an ideal spot for landing our submarine balloon." After towing *Trieste* all night, they arrived the next morning as a moderate sea was running. For once, all the dive preparations went without a problem, and father and son Piccard were soon on their way on what would become the world's deepest dive. Upon first leaving the surface they reported calmness and serenity as the vehicle was no longer tossed by the turbulent surface; then the light level depression gave way to "the last bluish gleams;" and finally there was the appearance of the phosphorescent animals. At 508 fathoms, they equaled the Beebe-Barton dive of 1934; next as their vehicle plummeted toward the bottom, they passed 748 fathoms, the depth Barton had reached in his benthoscope in 1949; and at 759 fathoms, they passed the depth mark of *FNRS-2* in its only deep dive. The Piccards felt it was a sort of chronological review of all previous dives. At 1,150 fathoms, the *Trieste* went beyond the depth achieved only a month earlier by pilots Houot and Willm of the French Navy in the *FNRS-3*, the vehicle Piccard had decided against working with. Now the Piccards were in virgin territory as they had been in their high altitude balloon ascents. The *Trieste* continued to descend at 3 feet per second until, nearing the bottom, some ballast was dropped to let them land gently. With a slight rocking, the bathyscaphe touched bottom at 1,732 fathoms (10,393 feet). The date was September 25, 1953. They spent only a short time on the bottom as surface weather was making up. The explorers decided to return to the surface immediately, and after dropping shot for some time broke away from the bottom. They returned to a rougher sea and decided to use the emergency compressed air to expel the water from the entry tower. It functioned well, throwing a stream of water out on the surface of the sea. On deck, Piccard noticed the the *Trieste* was covered with ballast pellet shot. He realized that due to the rate of descent and the turbulence caused by the bathyscaphe, they had exceeded the velocity of the shot and once on the bottom, the pellets already jettisoned had rained onto the deck.

From around the world, the Piccards received congratulations for achieving the deepest dive on record. But Auguste Piccard commented with some irriation, "However, that was not what I was after: the fact that the bathyscaphe had at last shown what it could do was enough for me."

The bathyscaphe was now a proven and viable device—it only needed some perfection to make it more versatile once on the bottom. Achieving this perfection was to be the work of the next twenty years of research. The culmination in the long history of man's attempts to explore in person the great depths of the oceans came some seven years after the first successful deep dive of *Trieste*, but it was the culmination by a definition based only on the depth achieved. It was hardly the end accomplishment —only the beginning. The age of iron-clad monsters, back-yard conversions, and heroic craft meant principally for the shallow waters was at a close. The year 1953 was the dawn of the true submersible and age of deep sea research.

# 4

# The Dawn of the
# Research Submersible

The *Trieste* is the real transitional character in our story of submersibles. It both closes the scene on the shallow vehicles and has an important part to play as a true full-depth submersible. Events that led to this fulfillment involve several Americans and the U.S. Navy, who had been watching with some interest the amazing accomplishments of the Piccards and *Trieste*, operating on an extremely low budget. These men were oceanographic scientists who had had a chance either to dive in the bathyscaphe or to associate with European colleagues who had. After the dives of 1953, Jacques Piccard had made only a few dives with *Trieste*, since the lack of adequate funds limited diving at great distances from port. One person to witness this operation and to appreciate the possibility of using the *Trieste* for significant scientific purposes was Dr. Robert Dietz, who was serving as the scientific representative of the U.S. Navy's Office of Naval Research (ONR) in London in 1956.

Dietz, a geological oceanographer, had chanced to meet Jacques Piccard at a conference in 1955 and became interested in *Trieste* for possible use by ONR scientists. After a visit to the city of Trieste, Italy in 1956, Dietz became excited over the potential of the *Trieste* to carry out sea exploration. Piccard, too, was excited because he had been trying for several years to find funds from the U.S. Navy for further diving. Piccard wrote of this incident: "I realized that line officers are primarily concerned with operational problems. It would take a marine scientist to appreciate a deep ship that could neither fight nor run fast, but merely sink to the bottom and rise again."

During the summer of 1957, ONR provided money to stage a series of fifteen dives in the Tyrrhenian Sea off Naples. It was primarily to allow American oceanographers to see how useful a bathyscaphe would be for exploration and making deep *in situ* oceanographic measurements. This first shakedown sponsored by the Navy showed that *Trieste* could carry specialized instruments such as hydrophones for sound measurements, could let the geologists inspect the bottom sediments, and allow biologists to see animals never seen by man before. By the end of that summer, *Trieste* made twenty-two dives and Piccard had convinced the ONR scientists that *Trieste* would be best utilized in the United States by the Navy.

In 1958, after negotiation with ONR, *Trieste* was sold to the Navy and transferred to the Naval Electronics Laboratory in San Diego, California. Along with *Trieste* came Giuseppe Buono, the Italian engineer, an expert who had been with *Trieste* since its birth. Dietz and Piccard both realized that they had a chance to put together a team that could assault the deepest depth of the oceans, a point located in the Marianas Trench, south of the island of Guam. But another team had already announced their intention of diving in the Challenger Deep—nearly 36,000 feet below sea level. This was the accomplished French Navy group led by Commandants Houot and Willm. Their success had come in 1954 when *FNRS-3*, which Papa Piccard had refused to build, had dived to 13,365 feet to capture the world record. The French Navy was determined to build a super bathyscaphe which could dive in the seven-mile-deep trench in the Pacific, but by 1958, when *Trieste* came to the United States, the French super-bathyscaphe was still two years away from completion. Dietz, eager to prepare *Trieste* to make the Challenger dive before the French boat, had selected a name for this ambitious undertaking—Project Nekton. Nekton is a term for the free-swimming animals of the sea, as opposed to plankton which refers to passive animals unable to make their own way and forced to drift in the sea.

If the first step in diving to the world's deepest place was bringing *Trieste* under the wing of the Navy, the second was in procuring a pressure sphere capable of 7-mile depths. Therefore a new sphere with a greater wall thickness was ordered from the famous German works of Krupp. It was formed in three sections, each 5 inches thick, and measured 6 feet in diameter. The old Terni sphere was safe to 20,000 feet and weighed 22,050 pounds; the Krupp sphere, designed for 36,000 feet, weighed 28,665 pounds. This required additional flotation, in the form of 6,000 gallons of gasoline, to be carried to offset the added weight. The float was lengthened to accommodate this extra gasoline. The new sphere arrived in San Diego in the spring of 1959. By mid-year, the Project Nekton team at NEL had acquired official Navy sanction and support. The diving team consisted of Dr. Andreas Rechnitzer, marine biologist, Dr. Robert Dietz, Lieutenant Don Walsh, Dr. Ken MacKenzie, and Dr. Jacques Piccard. Eleven men comprised the support crew.

*Trieste* departed San Diego aboard a ship bound for Guam. No mention was officially made of the planned plunge to the bottom of the world. Over the next two months, the *Trieste* team readied for the Deep Dive by making a series of increasingly deeper dives. Two of these were records in depth—they passed 18,000 feet on one and 24,000 feet on another. The main objective was focused on the Deep Dive, but on at least two of the seven Guam dives, Andy Rechnitzer was able to carry out biological observations briefly while they were on bottom at 18,150 feet.

Finally, when all was in readiness in mid-January, 1960, it was decided two days before the dive that *Trieste* would make the Nekton Dive in the Challenger Deep with Don Walsh and Andy Rechnitzer. Jacques Piccard found to his dismay that he had been scratched at the last moment from making this historic dive which he had been working toward for years. Luckily he had included a clause in his contract with the Navy that allowed him to participate "in dives with special problems." Diving safely into a trench less than a mile wide, under the prevailing heavy sea conditions, presented "special problems." Piccard won out and joined Don Walsh to make the dive on January 23.

While *Trieste* was remarkable in being able to dive to the deepest point in the oceans, it was not an ideal vehicle to take to sea nor was it suitable for much more than going straight down and back up to the surface. Getting to the site for Project Nekton was very slow going. The 200 miles to the site took over three days. Much of the time *Wandank*, the tug, wasn't able to make more than one knot punching into heavy head seas. Once the 600-foot tow cable broke. When they reached the approximate area of the enormous gaping trench, the support ship *Lewis*, a destroyer escort, determined the exact point of the deep trench by dropping nearly 800 dynamite charges through the night and recording the travel time of the sound to the bottom—just as an echo sounder does. They used dynamite to generate sufficient sound to reach to the very deep bottom and return.

The weather on January 23 was foreboding and stormy. The seas had built up overnight, some waves were now 25 feet high, making launching hazardous but not impossible. Piccard decided to dive. They had come too far to be turned back now. He and Walsh clambored aboard *Trieste* from a pitching, rolling motor whaleboat. They climbed down the long access hatch and prepared to dive. The Big Dive was on! The rest is history—*Trieste* plunged as rapidly as the pilots dared, and reached the bottom at 35,800 feet in a little less than five hours. The scientific results were startling to some who felt that abyssal trenches such as Challenger could support no life because of stagnant conditions with no water circulation. Piccard and Waslsh saw a small flounder move slowly away as they crowded, fascinated, at the view port of *Trieste*. The presence of active life was a valuable discovery, confirming the fact that currents, albeit slow ones, must bring oxygenated water from above to these great depths.

More than this scientific discovery, the Nekton dive opened the eyes of the world and marine community to the wonderful possibilities of diving submersibles as explorative tools. *Trieste* continued to dive for NEL on scientific missions involving acoustics, geology, biology, and physical oceanography. Yet she was not well suited to being much more than an "elevator". Biologists became frustrated when they couldn't stop and hover in midwater to observe plankton or small organisms. Other scientists became equally angry when the uniformed military officers made a majority of the dives for training and were not as excited about the phenomena of the deep sea as they were. Finally, after the nuclear submarine *Thresher* disaster in 1963, when 129 men were lost, *Trieste* was commandeered to Boston to participate in the submarine search. It was then clear to the marine scientific community in San Diego that if they were to pursue diving for science, it would have to be done from another vehicle, not *Trieste*. In August of 1963, *Trieste* had made 128 dives, and had been the catalyst that brought about the submersible era.

*Trieste*, over a period of ten years, had been heavily used and, in the eyes of the Navy, was in need of substantial modifications. The float that carried the gasoline was discarded, and the heavy

Krupp "ball" or sphere was placed in storage. The Terni sphere was mated to a more streamlined float for easier towing. The new ship became *Trieste II*. Thus *Trieste I* passed on quietly. It first lay sadly by the waterfront in San Diego, but was later placed in a small outdoor museum at the Navy Yard in Washington, D. C., and there was hope that the Smithsonian Institution might obtain it for display in its museum.

The design of its components set the way for nearly all the subsequent existing submersible technology. The use of acrylic ports and the 90° conical shape have been universal in all deep boats—unchanged since the design by Auguste Piccard. The configuration of the personnel sphere for deep vehicles—over 12,000 feet—remains the same as with the *Trieste*. Perhaps the greatest deviation has come in designing smaller, more maneuverable vehicles that can perform a variety of scientific tasks. *Trieste I* remains an outstanding prototype for all succeeding deeper vehicles which follow on the submersible tree.

The other significant type of submersible that has played a part in the development and increase of research vehicles was concerned with relatively shallow ancestors at about the time *Trieste I* was first diving. This development began as a logical extension of the aqualung or scuba diver, who was limited to about 250 feet in depth. Commandant Jacques-Yves Cousteau had started his trek into the mother sea in 1943 when he collaborated with Emil Gagnan in perfecting and patenting the aqualung. Cousteau and his group of colleagues aboard his research vessel *Calypso* carried out numerous explorations in the ocean—both by scuba divers and by devices suspended on long undependable cables. He also participated in the 1948 dive of *FNRS-2* off Dakar as Piccard's safety officer, and witnessed the birth of bathyscaphy. But, unlike Piccard, who took up immediately the ultimate challenge of the depths, Cousteau was more interested in extending man's capability in the sea beyond the meager 250 feet accessible with an aqualung. To be useful in the sea, he believed man must be on the spot to judge, evaluate, and carry out manipulative functions. Cousteau believed that cables and the devices that oceanographers hung on them to study the sea floor were doomed to do one of two things, "either become fouled or break." He firmly resolved in the early 1950s to carry out his exploration by "men on scientific submarines". Thus followed design work by l'Office Français de Recherches Sous-Marine (OFRS), one of Cousteau's several groups located in Marseilles, France. OFRS was then headed by André Laban, Director, and Jean Mollard, Chief Designer. Cousteau laid down the basic requirements of an underwater vehicle in 1955. The vehicle was to be primarily an extension of a scuba diver that would allow the underwater explorer to stay a longer time at greater depths, with the safety and comfort of being topside. Like a diver, the observer in this vehicle should have good visibility and be able to take photographs and small hand samples. Most important, Cousteau stressed, this two-man device must have maneuverability approaching that of a diver to allow close-in exploration.

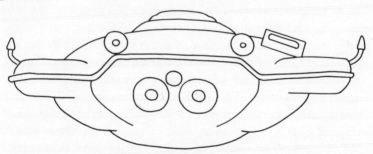

Cousteau *Soucoupe*

The result was the *Soucoupe Plongeante,* or *Diving Saucer,* which made its diving debut in 1959. The *Soucoupe* turned out to be all that its designers had hoped for. The personnel hull took the shape of a flattened sphere rather than a round ball. This allowed two observers to lie prone and look through two view-ing ports without the cabin having to be too large as in the case of a sphere. All machinery that could be was placed outside the hull to take advantage of the displacement in water and to cut down on weight. The hull measured 6½ feet at maximum diame-ter and was formed of two pieces of ¾-inch "mild" or common steel. A 15¾-inch hatch opening was at the top. Two openings were fitted with 6½-inch conical shaped acrylic ports for view-ing. An important feature, the designers thought, was a small port in the hull opening for the movie camera to view through. The safe operating depth of the *Saucer* was determined to be 300 meters, or 1,020 feet. Thus, among its various names, it became *SP-300*. Mollard's choice of 300 meters for an operational depth was most conservative and gave a 3-to-1 safety factor. The ballast system differed considerably from that of *Trieste*. *Soucoupe* was not a bathyscaphe, which had to achieve flotation for its heavy pressure hull by a float filled with gasoline. Designed only for the shallow depths of 1,000 feet, the hull was of light enough construction to be basically buoyant. Thus, by adding ballast weights that could be dropped, the *Soucoupe* entered the water slightly heavy, descended at about 60 feet per minute, and when the bottom was reached, one 55-pound iron weight was dropped. This action left *Soucoupe* with nearly neutral density. To ascend at the end of the dive, a second 55-pound iron weight was dropped. For small adjustments, a variable ballast system could add or drop several pounds by pumping water into or out of a tank in the cabin. This last system let the *Soucoupe* hover motionless in mid-water, not changing more than several feet over a fifteen-minute period. Such a capability allowed the study of mid-water animals not possible with *Trieste*. Further, the ability to scout in and around narrow canyons and caves offered exciting possibilities for the diving geologists.

When *Soucoupe* began diving for French scientists in 1960, little notice was taken of her, but shortly thereafter, in the fall, one of the NEL diving scientists who had been in *Trieste* had an opportunity to make several dives. He was Dr. Robert F. Dill, a marine geologist. Several years later, a fortunate meeting took place in which a vice-president of Westinghouse Electric Corporation, John Clotworthy, Manager of the Underseas Division, decided to work closely with Commandant Cousteau. Westinghouse had long been a manufacturer of torpedoes and sonar equipment for the U.S. Navy. Clotworthy saw the need to expand and diversify this military market, and believed building a small submersible was the way.

Cousteau agreed with Westinghouse in 1962 to design and construct a three-man submersible capable of operating to at least 3,000 meters, or 9,000 feet. By the end of 1962, a preliminary design acceptable to both Westinghouse and OFRS engineers resulted in a 12,000 foot, three-man vehicle called *SM-35*. Most of the components were modifications of the already successful *Diving Saucer*. The planned launching was set for early 1964. But by mid-1963 it was clear that there were problems developing with welding techniques, and the construction program would have to be delayed until these were solved. By that fall, the test welds on the Vasco Jet 90 steel hemisphere would not pass the notch sensitivity specifications required by Westinghouse, and completion of the vehicle, which the French and Westinghouse had named *Deepstar*, was rescheduled for 1965. A quick decision was made by Westinghouse engineers. Although it appeared possible to overcome the welding problems, further delays were unacceptable. Instead, the engineers took one of those trade-offs common to the engineering world. Westinghouse had, among its 105 divisions, a Marine Division that worked with a high-yield steel very acceptable to the Navy who certified submarines. This was HY-80, a 80,000 psi yield strength steel used in the construction of Polaris rocket launchers and in the hulls of all nuclear submarines. The Marine Division agreed to fabricate a spherical hull for *Deepstar* in four months. There was only one catch in this trade-off. The HY-80 steel hull to fit *Deepstar*, already designed, would be greatly de-rated in depth due to the lesser tensile strength of the steel. Instead of 12,000 feet in depth, *Deepstar* could now only safely reach 4,000 feet in manned operation.

Westinghouse officials were convinced that American oceanographers were eager to explore the regions adjacent to the coast. Several of these officials decided to test the diving waters with their big toes as it were. In early 1964, after agreement by Jacques Cousteau, and with the enthusiastic support of the Navy, Westinghouse paid to bring the *Diving Saucer* to California to demonstrate its general capabilities to the scientific community, to see if the submersible market really existed.

The OFRS team and the small yellow boat then nicknamed "Denise" came to La Jolla for a twenty-nine-day engagement during which more than fifty scientists from universities and Navy agencies all had a free demonstration ride in the canyons offshore at San Clemente Island. Scientists like Robert Dill of NEL helped to orient colleagues on ways to use the small *Saucer* for observations. The leased support craft was an 81-foot offshore boat with a small backhoe "crane" for launching the *Saucer*. Fitting everything on this small ship was a cramped arrangement but it worked. In many ways, it was a trial comparable to the ONR use of *Trieste* in 1957. Many eyes were opened to the amazing potential to explore places never thought possible even up to 1,000 feet. Following these dives the *Diving Saucer* was returned to its home in France. The explorers—like an underwater Lewis and Clark expedition—returned to tell other less fortunate colleagues of the remarkable things they had seen. Word spread fast. Key scientists like Dr. Fred Spiess, then Acting Director of Scripps, and Dr. Francis Shepard, noted marine geologist of Scripps, along with others who made dives, wrote several articles for scientific journals. This brought attention from funding sources in Washington. By last 1964 the Scripps and U.S.

Navy scientists had arranged with Westinghouse to recall the *Saucer* for a six-month period of dives. This was reasonable since *Deepstar* would not be completed for sea trials until mid-1965. Further, it gave Westinghouse a chance to assemble and train an operational team and support ship combination which would be ready when *Deepstar* came from France.

The budding of a new business began at this point. Westinghouse was in an ideal position to take full advantage of the enthusiasm of the scientist who was willing to loosen the purse strings of the ONR and National Science Foundation (NSF) to use the *Diving Saucer*. It meant paid learning for honest work. The timing was excellent. Although several other submersible ventures were in the works, none was ready to take to open water as deep as 1,000 feet in mid-1964. These others were, however, poised in the wings, about to make their first bow. They were *Alvin* and *Aluminaut*.

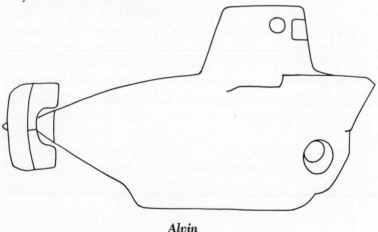

*Alvin*

*Alvin* was the result of the concept of Allyn Vine to provide Woods Hole Oceanographic Institution (WHOI) with a useful mid-depth vehicle, not unwieldy like *Trieste*, but better suited to scientific diving. The aims were close to those of Westinghouse.

WHOI had once considered using the all-aluminum submersible *Aluminaut* but had become hesitant when a $\frac{1}{16}$ scale model failed prematurely in a pressure test. Instead WHOI, with ONR funding, chose to have a vehicle built that would allow diving to 6,000 feet. The winner of the bid competition was General Mills, the breakfast food manufacturer, who had an electronics and manipulator division that successfully demonstrated they could build the three-man, 12-ton vehicle in slightly over one year. *Alvin* was completed in 1964—a year late. Before she was finished, Litton Industries purchased the division building *Alvin*. She was delivered and launched in Woods Hole in June, 1964, and only then did the real work of getting her fitted out and certified begin. Useful dives for science were still over a year away.

*Aluminaut* was also dipped ceremoniously in 1964, signifying about six years' work since its design began in 1958—over twenty years since the initial idea. J. Louis Reynolds of Reynolds Metals Company had long thought a deep submarine would be practical since he was an active champion of submarine cargo tankers during World War II. He had worked with Simon Lake on the design of submarines but had found no support for the project. Much later he chanced upon Dr. Edward Wenk who had long known that a deep submersible was quite possible to build. Wenk had performed a design study on pressure hulls at Southwest Research Institute in 1957, and Reynolds and Wenk teamed together to build a deep vehicle. The problem was to design a

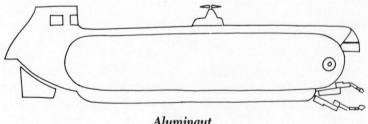

*Aluminaut*

# The Dawn of the Research Submersible

submarine strong enough to withstand pressures at 20,000 feet yet light enough to still be buoyant at that depth. The only materials which could meet this requirement were very high strength steels, titanium, and aluminum. Aluminum was chosen and the shape was an elongated cylinder to provide ample buoyancy. The original design goal of *Aluminaut* was 22,000 feet in depth. By the time preliminary design was completed, it appeared that the vehicle would probably operate safely to 18,000 feet. After scale model tests, this depth was reduced to 15,000 feet. The design called for eleven rings of 6½-inch thickness that had to be bolted together since welding that thickness of aluminum was not possible. Assembly of the submersible was carried out at Electric Boat Division of General Dynamics in Groton, Connecticut—builders of many of our fleet submarines and the company that John Holland started at the end of the 19th century.

*Aluminaut* was completed in late 1964, but further fitting out and sea trials lay ahead in 1965. While most of the world's smaller submersibles have their pressure hulls tested in a pressure tank, *Aluminaut* was far too large for any existing tank. Nevertheless, engineers are never happy until some indication of the strength of the particular hull has been obtained. In this case, strain gauges which show the movement and behavior of the metal were mounted inside the boat when it was tested to a depth of 6,250 feet.

In late 1964, there were several other shallow submersibles available but not in use on a regular basis. These were the several Perry *Cubmarines*—one that was rated to 150 feet and one to 300 feet. Electric Boat had recently completed a two-man vehicle for 600 feet, *Asherah*, for the University of Pennsylvania Museum to use for archeological work in Turkey.

But many other submersibles were in various stages of design. The aerospace industry was sniffing the wind and beginning to sense that there could be a market in underwater vehicles—perhaps not so large and dramatic a market as the space race,

77

but exploration of our own planet made better sense to many people who wondered what the real payoff was going to be in space.

The planned six-month *Diving Saucer* operation by Westinghouse moved ahead. A minimum of fifteen dives per month was contracted for by Scripps, to be shared by several Navy groups from California and Connecticut. The *Saucer* had been flown from France by a Military Air Transport Service 124-C Cargomaster—a tiny yellow creature lost in the great innards of this enormous aircraft. Diving began in November 1964, in the two submarine canyons off La Jolla. A 100-foot support ship was used to start with for local dives, but after one month a larger 136-foot offshore service vessel was found to be far roomier for the diving operation. A larger backhoe crane was placed aboard the stern, and a group of trailer living and repair vans used the 100- by 30-foot deck space. This same design was to be used for *Deepstar* and set the way for several other companies' subsequent diving services. By April 1965, the *Saucer*, her mother ship *Burch Tide*, and approximately eighteen crew members traversed from Santa Barbara to Cabo San Lucas, Baja Mexico, about 1,000 miles, making over 125 dives for about sixty different scientists on an almost daily basis. This diving operation was probably the first of its kind on such a long-term basis. It did a great deal to further convince the Westinghouse management and other major companies that the Navy and government-funded institutions had need for more and deeper-rated submersibles.

Further, many of the scientists who made the dives had become seasoned observers from submersibles. Also the *Deepstar* team was knit into a first-class operating group. The *Diving Saucer* since 1960 had completed over 450 dives for a total of nearly 1,300 hours in the water. This vehicle was over eight years old and going strong—a tribute to Cousteau and company.

The completion of this series of dives closes the transitional period of submersibles, which lasted from the initial dives of *Trieste* in 1954 through the *Saucer*'s remarkable diving service in 1965. These two vehicles and their respective inventor/builders are probably the two most important ancestors of the mass of second-generation submersible progeny.

Before we look in detail at some of these newer representatives and what they can do and have done, we should first consider the innermost workings of a typical submersible—what its component systems are and the way these strange craft are constructed.

# 5

# Submersible Components and Systems

A submersible capable of diving to continental shelves or the deepest oceans must protect the occupants from the external hydrostatic pressure exerted by the oceans. This pressure increases at a steady or linear rate—about ½ pound per square inch for each foot of depth. The exact amount for sea water is 0.445 pounds per square inch or, for a depth of 20,000 feet, the pressure exerted on a submersible hull would be 8,950 pounds per square inch, over four tons of force on each square inch of the hull.

The heart of the submersible is the pressure hull or cabin. The hull is the starting point in the design as well as in the construction of a vehicle. The characteristics of the hull determine the ultimate safe diving depth of the submersible. Often this depth limit is not determined until pressure tests are carried out after hull construction. A number of factors enter into the design, construction, and testing of the hull. Among these are the size, shape, and type of hull material together with the method of fabrication.

It is easy enough for an engineer to design a pressure vessel to take extremely high external pressures. Like Otis Barton with the Beebe bathysphere, one only needs to make the hull a bit thicker to resist the pressure. The real art in design is, of course, to provide maximum buoyancy with minimum weight. The term commonly used among engineers is strength-to-weight ratio. This simply means that as the submersible strives to go deeper, the hull must be capable of withstanding greater pressure. This in turn requires either greater thickness and thus greater weight, or a stronger material. So, the deeper we go, the greater becomes the problem—less buoyancy, less payload available, and greater strength required of the hull. A change in shape and stress points may also decrease ultimate depth.

The bathysphere of Barton and Beebe was far from being buoyant and had to be lowered on a cable—a dangerous proposition at best. Professor Piccard saw the solution in the use of gas bags or a gasoline-filled float to support the extremely heavy pressure cabin, thus eliminating the need of a cable tether. This was a truly ingenious approach for diving to the greatest depths of the ocean. But his type of vehicle had the limitations of heavy mass and immobility on the bottom.

The design of the hull is also determined by the vehicle's mission. The size of the hull is influenced largely by various components that will be carried either inside or outside the hull. A few of these components are: number of crew; life support equipment; instrumentation; scientific, navigational, and communication equipment; energy sources and conversion; propulsion system; pumps and hydraulics; ballast/buoyancy; manipulator tools. The weight of these component systems, their respective volume, and whether they are inside or outside, forward or aft, will bear directly on the hull design—the length, diameter, type of hull, and construction material.

The conventional submarine shape is an elongated cylinder, since the submarine began as a surface ship that occasionally dived. The long narrow hull made it fast on the surface and easy to dive by pitching forward. The shallow submersibles of the back-yard and simple machine shop variety continued to use this same shape—a cylinder with dome-shaped ends. In shallow vehicles, the common tendency is for the hull to be quite positive in buoyancy and to require ballast. This surplus buoyancy usually disappears around 2,000 feet. While this shape is practical for shallow ranges in the continental shelf depths, it requires reinforcing with ring stiffening for deeper depths.

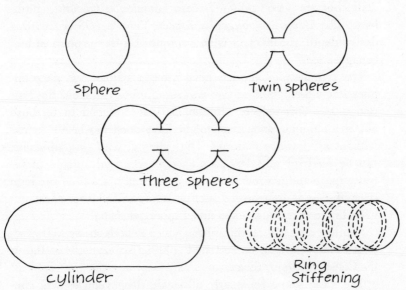

**Basic Submersible Shapes**

It was realized long ago that the sphere was the ideal shape, since the confining pressure is distributed equally on all sides.

Thus Barton used this shape for the *Bathysphere,* and his own *Benthoscope* which he built later. And Piccard naturally chose the sphere for his underwater balloon. If the sphere is an ideal solution, why don't all deep submersibles use it? Many have, of course, where small vehicles must go to 6,000 or to 20,000 feet. Yet the sphere presents complications in providing a large interior space. While the sphere-shaped hull is fine for a crew of four men, a sphere to accommodate eight or ten men is impractical. The inside shape of the sphere does not readily accommodate square instrument boxes. There is wasted space in fitting instruments to it unless an extra expense is involved in custom designing and modifying each rack-mounted piece to fit against the curved walls. Joining two or three spheres together has been done, but this, too, presents more complex engineering problems. The *Deep Submergence Rescue Vehicle (DSRV)* with a design depth of 5,000 feet is an example of a three spherical hull configuration.

The cylindrical shape has been used in at least one deep application, when a 50-foot size was required to carry out the mission goals. This vehicle, *Aluminaut,* was designed to be built out of aluminum since steel might have been too heavy to use without a buoyancy material. The cylinder, with ring stiffeners, can be used for mid-depth vehicles. It offers advantages of interior space and hydrodynamic shape. It is also a way to use high strength composite material such as reinforced plastics and laminates that can not be worked into a spherical shape.

Another variation on the sphere which is both strong and practical is the ellipsoid, or flattened sphere. An example of this is the Cousteau *Diving Saucer.*

The basic shape for nearly all small, deep vehicles will continue to be the efficient sphere, while the larger ones will use sphere combinations or cylinders. For some time, it is likely the common procedure will be to use two hemispheres forged, spun, and welded together. Experimental work has been done using

84

multi-faced, flat sides that are joined by a special cement, but this technique depends on the development of new structural materials.

Picking the right material for a pressure hull may sound simple, but it is really a complex matter in the case of the deep submersible. The available materials in use today and considered acceptable are steel and aluminum; to a lesser degree, titanium and acrylic are in experimental stages and growing in acceptance. Others with both promise and problems are glass, ceramic, and fiberglas reinforced plastic.

Steel is used in nearly every submarine and submersible in operation today, although it presents the designer with a variety of weight problems. The principal disadvantage of steel for a pressure hull is the low strength-to-weight ratio. This means that steels strong enough for pressure hulls are heavy and require supplemental buoyancy. Steels developed for the Navy submarines are classed as high yield strength (HY) and have been in use for over ten years. A typical one is HY-80, an 80,000 psi yield steel. While HY-80 is well suited for a fleet submarine operating to a thousand feet, it is not an ideal material for depths exceeding 5,000 feet. Steels with higher yield strengths are necessary as vehicles go deeper. HY-100, which designates 100,000 psi yield, has been used for at least one relatively shallow vehicle, *Star III*, at 2,000 feet, and for *Alvin,* which operates to 6,000 feet. In the search for better steels for deep submergence, much information has been gained from space technology where high strength and low weight were also mandatory. Steels that could withstand the stresses and heat of rocket launching were in some cases adaptable to pressure hull materials. Some of these range in strength as high as 300,000 psi. Few of them are thought suitable since there is the continued problem of welding the hull pieces together. It seems that as the yield strength increases in the more exotic steels, the welding problem does too, and there is difficulty obtaining sufficient weld wire and even heat distribution, as well as numerous other complexities discovered in using these steels.

The U.S. Navy, known for being conservative in selection and use of materials for deep submergence, has endorsed an HY-140 steel that was used for the *Deep Submergence Rescue Vehicle* (*DSRV*) launched in 1970. After working several years in the fabrication of this steel, the hull manufacturer, Sun Shipbuilding Corporation, determined that the yield strength was only 130,000 psi due to changes caused by welding.

But steel is not truly the answer because of its added weight. It only is suitable in the absence of a better, lighter, *and* more reliable material. Most of the present vehicles using steel and diving to 4,000 feet or greater must pay a heavy weight penalty by adding buoyancy. *Deepstar 20,000* using HY-130 steel will probably use 50% of its total weight for foam buoyancy—just to support the enormous weight of the steel pressure sphere. Other materials that are much lighter and, at the same time, much stronger, have been around for some time. Hopefully, one or more of these will be developed to a point of acceptability.

The use of aluminum seems logical because of its light weight yet relatively high strength. But aluminum used near seawater is readily corroded by salt. Such corrosion could become serious under pressure and cause metal failure. To overcome the process of stress corrosion, a thorough and elaborate coating of paint or anodyzing has been found a successful inhibitor of corrosion. Sinc 1960, aluminum has become quite acceptable for use in the pressure cases for deep cameras and lights, but relatively few applications have been made of aluminum for manned submersible hulls. The best-known example of aluminum for a submersible hull is *Aluminuat,* built of the toughest possible aluminum for Reynolds Aluminum Corporation. Aluminum welding is still a problem in joining hull pieces as it was when *Aluminaut* was built in 1964. Reynolds solved the corrosion problem by putting six coats of paint, each a different color, over the aluminum hull.

With careful inspection, it was possible to spot any chips or wearing away of the coating and thus prevent potential corrosion.

The Navy has been skeptical and slow to use aluminum for the problem of stress corrosion is the main drawback. Nevertheless, one experimental vehicle, *Moray*, was built by the Navy using two aluminum spheres. It was operated for about five years with no corrosion effects. No new construction is being undertaken in aluminum because there are other more promising materials. One of these is titanium.

Titanium is a relative newcomer in the underwater field of high strength, low weight metals. It is one of the "space age" metals used in constructing rockets. It has fabrication problems similar to aluminum in that it is difficult to weld and is also subject to stress corrosion. Experimental testing by the U.S. Navy in 1968 indicated that titanium was liable to fail catastrophically under conditions of stress in seawater if fatigue cracks or very sharp notches were imposed on test specimens.

The first applications of titanium were on *Alvin*. These new buoyancy spheres replaced the original aluminum ones at a saving of weight. Aluminum weighs approximately one-third more than titanium. Later, an experimental titanium hull was fabricated for *Alvin* which would extend its depth range from 6,000 to 12,000 feet by saving weight and gaining strength.

Recently interest has been shown in a variety of plastic materials. Among these are fiberglas reinforced plastic, fiber-wound plastic, and acrylic plastics. All of the plastics share the asset of being superior to steel and titanium in strength-to-weight ratio. Yet as with the previous materials, problems exist in the assembly and joining. Other areas that are not well understood are those of resin creep and the action of salt water on bonding agents.

The filament-wound plastic material has been highly successful in large rocket cases in aerospace applications. The highest strengths have been achieved in composite construction in which aligned, continuously wound filaments are embedded in a resinous matrix, usually of the epoxy type. Using this material in a continuously wound cylinder for deep submergence could produce a positively buoyant hull—a very desirable quality not found in steels. A look at filament-wound plastic, when compared with HY-150 steel, shows that the plastic has a 170,000 compression strength as opposed to 150,000 psi for the steel. But more importantly, where steel has a strength-to-weight ratio of $0.54 \times 10^{-6}$, filament-wound plastic has a $2.5 \times 10^{-6}$ ratio—nearly five times better. Experimental work on these materials has been undertaken at the Naval Applied Science Laboratory in Brooklyn.

Another Navy laboratory, the Civil Engineering Laboratory of Port Hueneme, California, has been at work for many years experimenting with different materials in seawater. Plexiglas has been used in deep submersible ports since Professor Piccard developed the conical port for *Trieste*. This material and his design has been used in virtually every vehicle since, with never a failure. Closely associated with plexiglas has been the clear acrylic sphere development for pressure hulls which allows the occupants unlimited vision. After tests using acrylic materials and bonding cements showed no degrading affects in over a year's immersion in seawater, the Navy commissioned the Southwest Research Institute to construct a 66-inch hull of twelve curved plastic pentagons, each 2½ inches thick. This hull became *Nemo*, one of the newer vehicles now being used by the Naval Underwater Center, and certainly an exciting possibility for scientists who want to look in all directions from one position.

The *Johnson-Sea-Link* vehicle launched January 1971 uses a 4-inch plastic hull that is said to be adequate for operations to 2,000 feet. As a diver lock-out vehicle it is intended for 1,000 feet. Two other similar acrylic hulls were initiated in 1970—both for shallow depths. This plastic shows greatest promise for the sunlit continental shelves, although future developments may extend the depth capability.

Another material that has intrigued designers is structural glass.

# Submersible Components and Systems

Several glass companies have been producing a glass ceramic material that has very high strength but at the same time is extremely brittle. For over ten years, small glass spheres of 9- and 12-inch diameters have been made for deep ocean applications exceeding 20,000 feet. Due to variations in the manufacture of glass, lack of quality control for flaws, and a general lack of understanding of the behavior of the material, these glass spheres are subject to random but catastrophic failure. Glass has severe joining problems. Hemisphere joining with clamps has been one solution with the small spheres. As the sphere becomes larger, the problems increase in making sure that the glass material is flawless. Prediction of material behavior and testing of each sphere is difficult. The U.S. Navy has officially rejected glass for any of its deep applications of manned vehicles. Yet experimental work with it is proceeding in shallow applications. Some materials experts say glass will never be acceptable—others are more optimistic as a result of the testing with small spheres.

This is a most brief summary of the materials being used through 1971 and those being developed for future deep submergence applications. The selection of high strength, ultra-light materials for the submersible pressure hull is one of the keys to more effective vehicles capable of greater ranges, speeds, endurance, and size.

Probably second only to the pressure hull in importance in the designing of a manned submersible is the life support system. The basic purpose of this system is to provide all occupants within the cabin with a safe, comfortable environment during the time of planned operation, plus an adequate safety factor. In simplest terms, life support should provide breathable air at about one atmosphere pressure, that which comes close to a "shirtsleeve environment."

The basic type which has been in use successfully for over ten years in vehicles such as Cousteau's *Diving Saucer* incorporates two fundamental elements. These are a supply of oxygen and a

89

means of removing the carbon dioxide discharged by the crew. In this way, oxygen ($O_2$) is added or "bled" into the cabin at approximately the same rate as the carbon dioxide ($CO_2$) is absorbed by a chemical "scrubber." In such an elementary system, the rate of $O_2$ flow is determined by the pilot who opens the

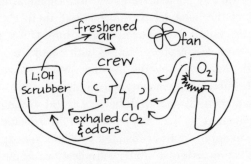

**Life Support System**

flowmeter valve connected to the $O_2$ tank. Since the respiration rates of persons vary, he must occasionally check that the $O_2$ and $CO_2$ rates are balanced. He does this by observing cabin pressure on a small barometer, being careful not to let pressure exceed one atmosphere (14.7 pounds per square inch) by more than one or two pounds. The scrubber in such a system is a chemical called baralyme or barium hydroxide, which has a strong affinity for $CO_2$. These chemical grains or pellets are placed in flat perforated trays that allow air to pass through. The pilot usually makes hourly checks on the $CO_2$ level, which is one of the gases that cannot be allowed to build up. One way he does this is by using a small breakable ampule with a colorimetric indicator. Air in the cabin is circulated by a fan to prevent pockets of $CO_2$ and to prevent the ports from fogging up. A spare bottle of $O_2$ and several containers of baralyme are always carried.

Most of the systems are designed to provide for several times the normal dive time in case of various emergency conditions.

# Submersible Components and Systems

The *Diving Saucer* has a maximum life support time of forty-eight hours for two persons. Other major vehicles have a capacity ranging from thirty-six to seventy-two hours. At least two have a life support duration of thirty days.

A number of improvements have been brought about in life support systems since the first Cousteau vehicle, although the elements are nearly the same. Most of the more sophisticated systems employ a regulator to control cabin pressure. This lets $O_2$ flow automatically from the storage bottle when needed. There are several $O_2$ regulators which sense the amount of $O_2$ in the air, and control the supply accordingly. Needless to say, these devices add to the cost. There are also systems that use a filtering and purifying apparatus to remove toxins and odors.

A very important component, although not technically part of the life support system, is the emergency breathing supply. The *Diving Saucer,* as an example of a basic minimum, had two mini-lungs for emergency breathing. These were merely 10-cubic-foot bottles with a regulator for breathing and a mask intended for use if an electrical fire or short circuit should make breathing impossible. This provided enough breathable air to last until the vehicle could be returned to the surface. A closed circuit system now in use is an improvement in which there is no unregulated discharge to the cabin and thus no possible buildup of pressure. Further, it has been found desirable to have a means of communication built into the emergency breathing masks so the submersible crew can easily communicate with the surface ship as well as with each other.

Other refinements that are necessary, both on longer missions and bigger boats, are humidity control and temperature regulation. Humidity is usually a problem in all submarines because much condensation forms inside as the vehicle cools down from the lower temperature of the surrounding waters. The more elaborate vehicles use air conditioning, as do our military submarines, not only for crew comfort but to protect the life and operation of the instruments which are seriously affected by moisture.

Perhaps the most highly developed life support system in a research submersible was designed for *Ben Franklin* for its 45-day drift mission. It used many of the latest power-saving techniques employed on space craft and supported five men comfortably for over thirty days.

Life support, while absolutely vital to the crew of the submersible, has not been a major problem to the design engineer.

The small research submersible is probably more plagued by the lack of adequate power than by any other factor. The development of electrical power sources suitable for deep submergence has not been nearly rapid enough. Our space craft use fuel cells, military submarines use nuclear power, and the commercial market is flooded with miracle batteries, but the average submersible struggles along with heavy, inefficient lead acid batteries. Why is this? In the case of the fuel cell and nuclear power, it is simply a matter of very high development costs. So far only the government has been able to pay for the development of submersible fuel cells, and the present fuel cell is regarded as impractical for existing submersibles because of its size and weight. Conventional batteries are readily available and, although the weight-to-power ratio is unfavorable—that is, high weight and low power—these batteries are dependable and relatively low in cost.

The earlier submersibles like *Diving Saucer, Deepstar,* and *Alvin* all used lead acid batteries, the same type as those used in automobiles. Batteries are relatively inexpensive to purchase, highly stable, and have a predictable lifetime. For this, the designer pays the price of lower output per pound of battery. The result is that a vehicle is limited to a relatively slow speed, a short range, and a fairly short dive time. In effect, power is limited severely. Most of the smaller vehicles are power limited. The *Diving Saucer,* with 14 kilowatt hours (KWH) of power was limited to two miles range at 0.6 knots, or a total of about four hours. Actually, most scientists admitted that four hours of intensive observation was enough.

However, in larger vehicles with more living space, there is an advantage to being able to "fly" at speeds of several knots for many hours while operating electronic equipment. To do this, vehicles such as *Aluminaut*, a 49-foot, 83-ton vehicle, use silver zinc batteries that have a much higher power-to-weight output ratio. This battery is considerably more expensive because of the cost of silver. Silver zinc is probably the most practical solution until further developments from fuel cells, oxygen/metal batteries, and ultra-small nuclear packages.

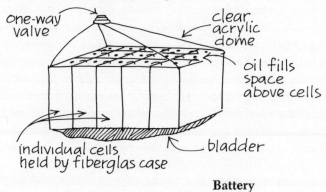

**Battery**

The general configuration of a lead acid battery for a submersible is slightly different from what we have in our automobiles. Instead of the entire battery being sealed in a case, the plates of each cell are put in a jar. These jars are placed in a large rectangular case. Sufficient electrolyte is poured in to cover the plates and a nonconducting oil is poured on top. The oil does not combine with the electrolyte but fills all the empty space in the case top. Finally, at the peak of the case top, there is a one-way escape valve. This cap allows the gas generated during battery use to flow out, but prohibits seawater from entering.

93

Beneath the case is a neoprene diaphragm or bladder, which is also filled with the same electrolyte. When the seawater pressure presses on the battery there is always fluid to fill the battery case completely, thus insuring that there will be no void spaces. Any gas discharging from the cells merely bubbles out the escape valves.

This is called a pressure-compensated system, as opposed to placing the batteries in a pressure-proof container, and the battery in this instance can be used at any depth since it does not have an external pressure case. It has been the practice of designers to place all weights such as batteries outside the hull, and thereby gain the advantage of the displacement in sea water. As a slight benefit for the designer, the battery can be used as ballast by placing it below the center of gravity. In many cases, the batteries have been designed so that they can be released to obtain buoyancy in an emergency. Further, lead acid batteries are generally considered safer placed outside where any outgassing or discharge of hydrogen gas will be harmless. Several vehicles place silver zinc batteries inside, where vehicle buoyancy is not critical. *Aluminaut,* for example, has adequate buoyancy by virtue of a 45-foot pressure hull seven feet in diameter. She carries several tons of silver zinc main batteries inside.

There are several voltages used by submersibles. The two most common are 120 volts and 24 volts. The U.S. Navy has long believed in low voltage in seawater and stays with 24 volts wherever possible. However, there are certain electrical efficiencies in using a higher voltage, so a number of designers have chosen 120 volts.

At the present stage of development, batteries appear to be the only dependable and affordable source of power. The first and only experimental nuclear reactor research submarine, *NR-1,* has been in operation only a short time. It cost nearly $100 million and is over 130 feet in length. It is clear that few private groups or companies could afford this type of power.

The purpose of the electrical system is to distribute the electrical energy from the power source for the functions of the vehicle, namely propulsion, lighting, and instrumentation.

In most submersibles where the batteries are external, the largest electrical drains are for propulsion motors and incandescent lighting. There has been a problem involved in passing larger amperage wires through the hull with the use of standard penetrators. Instead of running the wires inside to a switch or controller, the practice has been to use sets of contactors or relays housed in oil, and mounted outside. These act as the switches for large current flow, while a small diameter wire is passed through the hull to turn the contractors on and off. Yet most operators of vehicles agree that contactor boxes present maintenance problems.

While the early vehicles such as *Soucoupe* and *Trieste* used only direct current (DC), later ones incorporated AC inverters into the electrical system. Scientists wanted to operate some of their equipment on alternating current (AC); more efficient lighting was possible from AC; and better, lighter propulsion motors could operate on AC. So second generation vehicles like *Deepstar 4000* were designed to provide enough 120 volt AC for propulsion. This was done by using a rotary inverter which changes DC to AC. The rotary inverter mounted outside was a heavy, bulky device similar to a power generator. After several years of use, Westinghouse Electric Corporation, owner of *Deepstar 4000*, installed a "static" or solid-state inverter. This substitution saved much weight and space, although it was less reliable when first installed. Later modifications corrected this situation.

The distribution of electricity is made inside the vehicle. Power from outside, whether converted to AC or from the DC supplies of batteries, must pass through the hull by what is called a penetrator. This is a plug in the hull that usually allows a number of wires to pass through with no danger of any water entering. The wires entering are generally small, since they carry very little power and are used for controlling outside functions or for passing signals from outside sensors such as sonar, temperature probes, and current meters. All these wires—and there may be several hundred on the medium-sized boats—are routed to a distribution panel. The power levels can be monitored here, fuses and safety

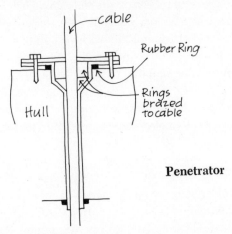

**Penetrator**

breakers are inserted, and if needed in the more elaborate vehicle, alarms and emergency switches are added. The panel or panels, depending on cabin layout, are usually situated within easy glance and reach of the pilot and copilot.

One of the sources of considerable trouble with nearly all submersibles has been the cable connectors. The purpose of a connector is to join the cable from a particular instrument or sensor to the penetrator and thence inside to the cabin. If a sensor, for instance a sonar transducer, has to be removed for replacement or recalibration, it is easier to unplug it than to cut a cable and then resplice it.

Many of the first connectors were adaptations of wiring cable connectors used on land with caps or seals over them. The increased problems of salt around low-level electrical signals were not solved by most of these. The rough handling and generally poor conditions in which connectors were taken apart at sea

made failure rates unacceptably high. Several companies under-
took special designs to improve the connector. One of these improvements resulted from developmental work done by the U.S.
Navy Electronics Laboratory in San Diego with the *Trieste*. There
had been a need for a connector that could be joined or mated
while underwater. The *Trieste,* when on a dive mission, was
towed to the site and left in the water. Therefore, any changes in
cameras, lights, transducers, etc. had to be made underwater. A

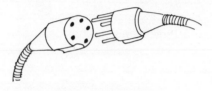

**Self-purging Connector**

connector was developed that allowed this. It was called "self-
purging," because as the pin was inserted into the socket, the
water was forced out and the surface wiped clean enough to
allow contact of conducting members. This connector was later
manufactured commercially and used on many submersibles
where underwater connectors were necessary.

At a time when connector failure rates were high, several
submersible groups chose to splice wires, eliminating a connector,
rather than taking a chance that a connector would fail, causing
instrumentation malfunction. More attention to quality control
and improvement of design has brought better, and consequently
more expensive, connectors on the market.

Nearly all submersible propulsion has used propellers driven
by electric motors, but some of the exceptions deserve comment.
The *Soucoupe* of Cousteau, designed by Jean Mollard, used
water jets driven by a water pump. This gave *Soucoupe* high
maneuverability, although slow speed. The conversion of electric
motor power to water pump was quite inefficient, but was a sat-

isfactory type of propulsion for that vehicle. Another type used experimentally on a U.S. Navy submarine, the *SS-XI*, built during the 1950's, was hydrogen peroxide. It served the purpose of being ultra quiet and still efficient, but the application had not been perfected. An explosion with the system in 1958 discouraged the Navy and hydrogen peroxide was abandoned.

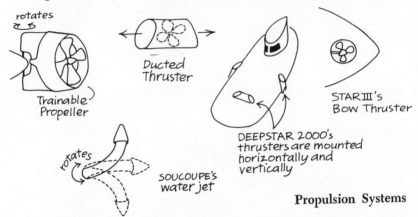

rotates

Trainable Propeller

Ducted Thruster

STAR III's Bow Thruster

rotates

SOUCOUPE'S water jet

DEEPSTAR 2000's thrusters are mounted horizontally and vertically

**Propulsion Systems**

The requirements for a propulsion system of a research submersible are mostly for small weight and volume with relatively high efficiency. The noise characteristics are of less importance than in vehicles with a military mission. Generally, a propulsion unit consists of a motor with a propeller connected directly to it, and a controller. Some systems have used hydraulics to drive the propeller. Many vehicles have used two propellers, which give increased maneuverability.

There are probably as many varieties of propulsion systems as there are submersibles at this stage of development. *Alvin*, for example, has one stern 7.5-HP hydraulic motor driving a 4-foot diameter propeller that is trainable—it pivots in the horizontal plane for steering. There are two 2-HP motors mounted midships, one port and one starboard. These motors can be tilted in the vertical plane so that the propellers will drive the vehicle up or down.

Lateral thrusters placed at the bow or stern have been tried on vehicles such as *DSRV* and *Deepstar 2000*. These are usually small motors that drive a propeller ducted in a sort of tunnel, forcing the bow or stern to move laterally. Such a thruster allows, at very low forward velocity, a quick change of direction not possible with the main propeller and rudder.

The major controversy over propulsion systems is and has been for some time, whether they should be DC or AC. There seems to be no simple answer to this question. Of course, the question of AC and DC is hardly new. This battle also raged at the turn of the century, as America was beginning to introduce electrical power into the major cities and a few homes. AC, of course, has won out for most consumer use—in spite of the predicted dangers of AC being a safety hazard.

DC is certainly a simpler means for propulsion, since no conversion from the battery power is necessary. In the mid-fifties, there was little thought of using AC motors with conventional invertors. DC motors can be run directly from the battery supply with good speed control and relatively small expense. But as DC propulsion systems were used on submersibles, a number of problems developed. There are several problems with most existing DC systems. DC motors have commutators and brushes and although operated in oil-filled cases they are very sensitive to any salt water intrusion and may short circuit. Many motors require maintenance of the brushes every forty or fifty hours of operation. Further, a DC motor itself is considerably heavier than an AC motor of comparable horsepower.

The AC induction motor is a simpler construction since it does not use a brush and commutator and is therefore lighter. But to run it, the DC from batteries must be converted to AC and the weight of this equipment may equal the amount saved on the AC motor. Further, experience on *Deepstar 4000* showed Westinghouse that the early static inverter was plagued with failures.

But it is possible that with a more dependable static inverter, AC will better suit very deep-diving vehicles. Westinghouse was planning to use AC for its *Deepstar 20,000*. But it is probably too early to commend or condemn one or the other as both require further developmental work.

One of the common beliefs aboard a deep submarine is that to be effective, the propulsion system should give it moderately high speed. But, for nearly every application near the bottom, speed is not needed. Most vehicles boast a maximum speed of three or four knots, but few use this rate because it rapidly drains power and observing scientists are usually content with a speed of one knot or less. For long search missions, there may be justification for greater speeds, but not greater than 4 or 5 knots. Most of the small boats like *Soucoupe* or *Alvin* are not designed for speed, as a quick look at their shape will reveal. Others, with a much greater length-to-diameter ratio (L/D) are able to have higher speeds. *Moray*, for example, measures 33 feet by 5 feet in diameter. Her advertised speed is 16 knots—probably the fastest of all the submersibles. Her mission was one of testing sonar equipment, rather than visual observation.

Propulsion continues to be closely tied to power souce development. High speed, long endurance systems requiring above 100 HP will only come about with the development of much better power sources such as fuel cells and nuclear packages.

One of the more interesting developments in propulsion that may have future promise is the Varivec by Westinghouse. This system uses a large number of small blades mounted radially on a ring. The blades can be controlled in collective pitch for ahead and astern thrust as well as in a cyclic pitch mode where up/down-left/right thrust vectors can be generated. This propeller can maneuver a vehicle without additional rudders or stern thrusters.

The ballast system of a submersible, like that of a submarine, is responsible for diving, surfacing, as well as trimming for operation at specific depths. The conventional submarine goes through a procedure of flooding ballast tanks to change buoyancy from positive to negative. To return to the surface, high pressure air is blown into these tanks, expelling water and giving added buoyancy. Variations of this technique are used in the shallow submersibles. Below several thousand feet it becomes difficult to use compressed air for blowing ballast because of increased pressure. Some different approaches were adapted from the design of the *Trieste*.

*Trieste* was operated very much like a balloon, using iron shot as droppable ballast and gasoline from the float to reduce positive buoyancy. In this way it was possible to regulate the ascent and descent of the bathyscaphe.

Cousteau changed this in the *Soucoupe* by having two cast-iron ballast weights of 55 pounds each. With both in place at launch, the vehicle was about 50 pounds heavy and sank to the bottom at about 60 feet per minute. Near the bottom, or wherever it was they wished to operate in mid-water, the pilot would release the first of these weights, making the *Soucoupe* nearly neutral—that is, neither heavy nor light. Usually, a slight ballast adjustment is necessary at this point. An important part of any ballast system is used next—the variable ballast pump. Scientists have found it one of the most valuable features in making small depth changes. The *Soucoupe* had an internally located pump and a water tank. Water could be admitted through a pressure-reducing block to this tank, thereby making the vehicle heavier. Or, the pump could expel some water, making *Soucoupe* lighter. This admittance of outside water at a maximum pressure of 445 pounds as it passes through the hull is permissible at 1,000 feet, but is viewed as a potential failure point on deeper vehicles. Therefore, all deep boats place their variable ballast tanks outside.

The type developed by *Alvin* has been adopted by various other boats. It uses oil in a closed system that is pumped from

# Diving for Science

titanium spheres to collapsible rubber bags and back again—all outside the vehicle. When the oil is pumped into the bags, the amount of seawater displaced by the vehicle is increased, thus increasing the buoyancy. *Alvin* then rises slightly. Pumping oil back into the spheres makes the vehicle heavier. This system

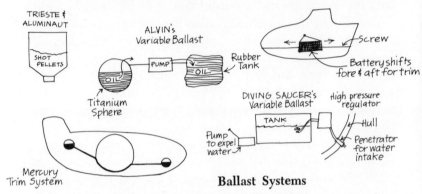

**Ballast Systems**

allows for 500-pound changes permitting additions, differences in weight of personnel, and changes in density of seawater. Using a pump with fine control permits making very small adjustments in weight to compensate for dropping markers or taking on core samples. To ascend, the *Soucoupe* pilot released a second 55-pound weight. If extra buoyancy were needed in some emergency, a 400-pound lead weight could also be released.

Another part of the ballast system is the fore-and-aft trim that gives a vehicle a bow up or down angle. One of the simpler techniques for doing this is used by *Pisces I*, whereby the battery case was moved forward or aft of a center point by a worm drive. A more sophisticated approach was designed into *Soucoupe*, using an expensive, heavy commodity of mercury which can be pumped fore and aft as liquid. A tank is located in the bow and another in the stern. Each is half full of liquid mercury which is pumped quickly from one to the other, giving a 30° bow angle up or down. This allows observers to have a better view

102

and to be able to aim cameras at the field of interest.

Another factor to consider for a deeper vehicle is the method of gaining flotation. Most spherical-hulled vehicles designed for depths of 4,000 feet or deeper require more buoyancy than the hull offers. One of the best solutions for this problem has been the use of syntactic foam—the microbubbles of glass and epoxy resin that can be cut in many shapes to fill void spaces. This foam, in block form, has a density ranging from 42 pounds per cubic foot to 30 pounds per cubic foot. Foam of higher density absorbs almost no water and has been pressure tested to 20,000 feet. The flotation it offers is not great, since the vehicle achieves a pound of buoyancy for each half pound of weight. While this is a high price to pay, the only alternatives are the use of spheres of titanium or floats of gasoline, neither of which is as suitable or as safe as syntactic foam, which has a great advantage in that it is so easy to fit into the small spaces where the shape of a sphere is inefficient.

In the deep vehicle, the hull of which is made of thick-walled steel, a large portion of the total vehicle weight is devoted to foam flotation. *Deepstar 20,000* proposed to use 20 tons of syntactic foam to support the nine-ton pressure hull for an over-all vehicle weight of 42 tons.

For a number of vehicles, hydraulics have been chosen to drive propellers and operate manipulators, drills, light booms, etc. The advantage of hydraulic activation is that the pump can be placed outside, and using electrical control, a number of functions can be efficiently performed. However there are some disadvantages. Hydraulic systems have a habit of leaking and creating a mess on deck. If seawater gets into the hydraulic piping system, it can do corrosive damage and consequently there may be frequent maintenance problems.

A simple hydraulic system has an oil reservoir, a pump that maintains a pressure in the system, and a pipe system that leads to various actuators. The *Soucoupe* probably had one of the simplest systems. In this case, it was easier to mount the pump and reservoir in the cabin. From there, pipes went through valves and a penetrator to the actuators outside. The pump kept a positive pressure in the system at all times that was higher than the seawater pressure. In this way, if any leakage occurred, it would be of the oil to the sea and not the reverse. Most systems have followed this basic design.

While many operators may consider hydraulics a nuisance, hydraulic power is more efficient than electric power for most functions. Each actuator is far lighter than an electric motor of the same capacity. It is a design advantage to use hydraulics in a larger vehicle, whereas the very small, light vehicles such as *Nekton* and *Cubmarine* rarely use any hydraulics and would not benefit from such a system.

The means by which vehicles are controlled and instrumented varies vastly from one submersible to another. The simplest design of a submersible uses a rudder mounted aft of a main propeller and some sort of diving planes near the bow for control surfaces. An example of this simple system is the Perry *Cubmarine*. The pilot has a stick or wheel that controls both rudder and planes. He also has a throttle which may be as rudimentary as Off, Half, and Full Speed. Other instruments show heading or direction, depth, and battery power remaining. There is limited maneuvering control with such a system. At low speeds of less than one knot it is nearly impossible to steer and turn sharply since there is not enough water flow past the rudder or diving planes.

An unconventional improvement on this sort of control appears in the *Soucoupe*. Its propulsion system uses water jets located on the forward wings. These can be rotated, one fore, one aft, for turning, or, with the rudder, all water flow can be diverted into one jet. Control of these functions is by hydraulics. The pilot can also control the vehicle's attitude by pumping mercury fore or aft for bow down or up. There are no control surfaces as in the *Cubmarine* and high maneuverability can be achieved at low speeds. To "fly" *Soucoupe,* the pilot lies prone on a couch, facing a port. This permits him to see with ease and control the vehicle at the same time. The observer lies next to the pilot, using his own port which has an overlapping view with the pilot's port. Relatively few instruments showing the status of vehicle functions are used in a simple vehicle like the *Soucoupe.* The pilot has the directional gyro and depth gauge in front of him. Other displays, such as echo sounder and the operating instruments, are to either side. All *Deepstar* vehicles follow the Cousteau principle of putting the pilot and observer with their faces to a port, since seeing is a major mission in submersibles. Thus the pilot flies his vehicle visually most of the time.

Another type of submersible control and instrumentation is used in larger, more complex vehicles. An example is *Deep Quest* in which pilot and copilot sit upright, side by side, each at an instrument console. While they have port viewing, it is not as accessible as the face-to-port arrangement, and the vehicle's control is not dependent on visual contact. The sit-up school of instrument layout and controls is closely allied to the aircraft and submarine experience of designers. In *Deep Quest,* designed and built by Lockheed Missile and Space Corporation, the pilot has a very complex set of instruments and gauges, compared with most submersibles. He has a set of commands that the vehicle can be given, and with the use of a small computer similar to those on aircraft, the vehicle will navigate itself to a requested location. It can, of course, be flown manually by the pilot. A television camera outside gives the pilot a view on his console monitor. This vehicle and its second generation child, the *DSRV,* are probably among the most sophisticated submersibles ever built.

One of the control features that has been adopted for many vehicles was developed for *Alvin* by her ingenious crew at Woods Hole Oceanographic Institution. This is a portable control box that has propulsion controls and indicators, which permits the pilot to move around inside near the port. It also allows vehicle control from the top of *Alvin*'s conning tower, which is required as part of its docking routine with the mother ship.

After about five years of using some twenty research submersibles, there is general agreement by operators on vehicle instrumentation requirements, many of which are involved with navigation. A submersible vehicle should have the following minimum equipment:

1. Compass for heading or direction—preferably a gyrocompass, such as a MK-27 gyrocompass and a magnetic compass as a back-up.
2. Depth gauge of high accuracy—less than one percent full scale. Most vehicles have several gauges with different accuracies.
3. Speed measuring instrument—current meter or flow meter.
4. Communication system—an underwater telephone using an acoustic signal such as the standard Navy (UQC)frequency; a surface radio with 10-20 mile range, FM preferred; and hard wire for predive checks on surface.
5. Forward scanning sonar for obstacle avoidance—one of the most important tools in search navigation. Most vehicles have Continuous Transmission Frequency Modulation (CTFM) type.
6. Echo sounder—preferably with up and down transducer for measuring distance to surface and bottom.
7. Voice Recorder/Log—a small tape recorder with capacity for entire dive.
8. Xenon flasher for surface identification visible for 2 to 5 miles.
9. General lighting—incandescent and mercury or thallium iodide for operation on the bottom.
10. Timer/Chronometer—a multiple-hand timer.

11. Pinger or transponder for surface tracking and emergency signalling.

Not all of the submersibles in use have all the instruments listed since some of the smaller ones lack the power supply or extra buoyancy to carry such equipment. For example, many, such as *Cubmarine, Nekton,* and *Submaray* do not have sufficient battery reserve or interior space for a CTFM forward scan sonar. Most operating groups feel that this sonar is needed for accurate navigation and to allow the vehicle to maneuver into confined areas of interest. With this equipment, there is a better chance of locating bottom objects and making significant scientific observations. Many an expedition has gone awry when it found that the positioning of the submersible was grossly inaccurate, and the dive survey was for naught.

The most important equipment for the scientists is his own gear and that supplementing it. This usually is called the mission-oriented equipment.

It is easy to see that there can be a great variety of specialized equipment each scientist may have developed for his particular series of dives; yet there are a number of similar requirements shared by most investigators. These can be best divided by scientific disciplines as well as a general category.

The quality and reliability of cameras is of utmost importance, since 85 to 95 percent of most geologic and biologic dives are documented by photographs. Some of the submersibles have been thoughtfully designed almost entirely around the camera. The Cousteau-inspired series—such as *Soucoupe, Deepstar 4000, 2000,* and *20,000*—all have a movie camera looking through a central port located between the two forward ports. Thus, whatever object the observer or pilot sees can be recorded on film at the moment it is seen. And by having the camera inside, new film cartridges can be installed easily. Some vehicles such as *Alvin* were designed without a movie camera port. When a scientist desired motion pictures, he had to hand-hold a camera at one of the ports, a practice which resulted in poor photography and the risk of scratching the port. Putting a movie camera outside is possible, but with a 400-foot film load for a 6,000-foot depth, the camera pressure case becomes unwieldy. On the new Cousteau minisubs, the 16mm cameras are outside, since the vehicles go to only 2,000 feet, and thus a smaller camera case can be used.

A still camera, on the other hand, is suitable for external mounting since it is much smaller than a movie camera, and with a total of 400 exposures, is easily included in a case 7 inches in diameter and 24 inches in length. The camera is controlled from within and may have refinements such as zoom focus, frame counter, and a data chamber that records date, time, tilt angle, and water depth on the film. Color film is usually used for animal identification and geologic samples. These cameras use either 35mm or 70mm film. Several subs have employed a pair of still cameras for stereo photographs, thereby making possible a good indication of size of objects and animals. With stereo or single cameras and the necessary lighting, the scientist can produce a basic record of all he has seen on a dive. If a camera malfunctions, jams, or a strobe does not fire, the record of a major portion of the dive could be lost. This sort of thing has happened from time to time, spoiling the anticipated results of a dive that may have cost the scientist $10,000 of his limited research budget.

The inside shape of submersibles varies from spherical to cylindrical. Most of the research vehicles have been modified to provide several cubic feet of standard rack (19 inches wide) space so that there is a place to mount such things as special tape records, amplifiers, and data systems of the different scientific parties that come aboard.

At the time of design, several of the DRV designers plan the inclusion of different power supplies for the scientific investigator. Most boats supply raw, unregulated 120 volts DC or 12 volts DC.

By including power conversion as part of the submersible system, there will be 12 volt DC, 24 or 28 VDC, and 110 VAC available in small quantity for specialized instrumentation. With this, a scientist may be able to bring a piece of laboratory equipment aboard without rebuilding it or providing a special transformer or inverter. This is important when an investigator may only make one or two dives and would not otherwise need a special adapter.

Several vehicles have had a "brow" or lightweight framework outside which permits the quick mounting and detaching of heavy equipment such as cameras and lights. The brow is made of aluminum tubing, fastened at points on the boat that are strong enough to carry heavy loads. The framework may extend 3 to 4 feet forward of the vessel, placing the lights as far as possible from the view ports to prevent light backscattering by particles in the water.

The geologist, while relying heavily on movie and still photos, also has devised a number of other ways to investigate the sediments, rocks, and general terrain of the bottom. The most universal of these is a way to take samples. Usually the geologist has used the vehicle's mechanical arm or manipulator to pick up sediments or rock samples and drop them into a bag or compartment. The scientific purposes for taking such samples will be discussed in Chapter VI.

Several submersibles have had specialized corers built that can take multiple piston cores of soft sediment up to 6 feet in length. This is merely a hydraulic ram that pushes the 2-inch plastic tube into the bottom by using a cutting nose and core retainer. An example is the one used on *Deep Quest*. The lighter vehicles have difficulty carrying such a large device and lack the hydraulics or power to operate it. *Alvin* has a smaller corer which works well. *Soucoupe* and *Deepstar* use a coring tube about 24 inches long held by the arm which punches the tube into the sediments by shifting mercury trim forward. Sometimes pilots found that additional penetration, and thus a larger core, was possible when all crew and passengers inside lunged forward. The larger vehicles are better suited to obtaining geologic cores because of greater weight, stability, and power.

At least three vehicles, *Alvin, Archimede,* and *Nekton,* have hard-rock coring tools designed to get several feet of rock core for examination. *Nekton,* although a small, inexpensive, and shallow diving vehicle, was designed by a group of highly-experienced diving geologists who had gained experience on many other submersibles. The rock core/cutter is rigged so that cuttings can be collected through a lock that permits them to be brought inside the hull for immediate inspection by the geologist. The ability to penetrate a thin coral growth and sample the rock beneath is of great value to the geologist.

A method involving indirect sampling, is a sub-bottom profiler. This acoustic device has been often used on surface oceanographic vessels. By transmitting a high-energy signal from a transducer into the bottom, it is possible to see, on a paper recorder, layers reflected by the sound, indicating the top sediments and underlying structure of rocks. Profilers used on submersibles are low-powered, but do, nevertheless, obtain recordings with up to 50 feet of penetration into softer silt sediments. Sound penetration in sand is limited to about 10 feet.

Using acoustic energy as does the sub-bottom profiler, but energy of a different frequency, a special type of sonar has been developed to look to each side of the vehicle for purposes of searching for objects and for bottom-mapping. This sonar, called a side-looking sonar, sends out a narrow fan-shaped beam on either side of the sub. The reflections or "high-lights" show up on a paper recorder to form a map of all bottom features. It has been used on several larger vehicles which can carry the weight and also have the required stability.

The biologist needs to record the activities of marine animals, the number or population density, and the sizes of organisms.

He also may wish to collect some of these organisms. At times, several square-mouth nets are mounted on the bow of a vehicle with closing releases controlled from inside to net small slow-moving organisms. Such nets were used on *Star III and Alvin.*

A giant "slurp gun" was built for *Soucoupe* and *Deepstar 4000* by the Navy Underwater Center to suck in small creatures by a pump. This big "vacuum cleaner" consisted of a pump housed in a cylinder and a hose with a funnel-shaped end attached to the manipulator arm. On several occasions, it captured rare species of fish which had previously evaded nets and divers. Part of the success was due to the use of an anaesthetizing solution which temporarily stunned the animals and let the funnel-shaped scoop "slurp" them into the collection bag.

*Deep Diver, Shelf Diver, Beaver IV,* and *Sea-Link* are equipped with a bottom hatch to let divers, supported on an air or gas mixture, swim out to collect specimens and then return. This feature, although little used to date, may turn out to be extremely useful to biologists for Continental Shelf diving. Other new vehicles can be expected to have this lock-out feature.

Physical and acoustic investigations depend on instruments to make measurements in the water column and near the bottom. Several versions of an automated system using electrical sensors and recording equipment have been designed and built especially for the small submersible. This instrument is called an S.T.D., for Salinity-Temperature-Depth. It allows the physical oceanographer to describe any one point in a water mass by characterizing its salinity and temperature at any depth. The best and most sophisticated system so far was designed for *Alvin* in 1968; it records digitally all information on magnetic tape.

While most vehicles have their own speed or current meters, few of them are very accurate. Further, meters mounted near the vehicle may pick up turbulence created by the vehicle itself. *Deepstar 4000* was equipped to use a special Savonius meter that could be planted on the bottom while the vehicle slowly

111

backed away. A 50-foot cable connected the meter to the vehicle.

A number of different size and different frequency arrays have been mounted on several vehicles beginning with *Soucoupe*. These arrays are large, cumbersome frameworks or "booms" on which small hydrophones are mounted. Signals received by each hydrophone are wired to the cabin. Inside, the scientist may have a large bank of amplifiers and special tape recorders where all the sounds from the ocean are stored for analysis later.

The manipulator is a device that allows the persons inside to extend themselves into the outside ocean environment to carry on functions of gathering, welding, cutting, and a long list of manipulative functions. Many times a scientist may have to choose whether he carries the manipulator, which may weigh as much as 500 pounds, or whether he leaves it behind in order to carry other mission-oriented scientific gear, since the payload allowance will not permit both. The evolution of the manipulator started with the *Soucoupe* and *Trieste I*. *Soucoupe* carried an elementary arm made of stainless steel, with a clam-shell claw at the end. The arm was controlled hydraulically by the pilot who could pick up small objects with it and place them in a basket on the vehicle. The arm had a single motion, pivoting at the shoulder. To reach down, the pilot tipped *Soucoupe* forward and down with the mercury trim, until the claw was able to close around the desired object.

*Trieste* had an articulating arm with shoulder, elbow, and

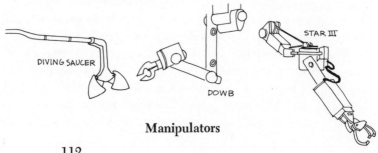

**Manipulators**

112

wrist motions. It apparently worked, after a fashion, but during the *Thresher* search, it developed "arthritis," or pronounced stiffness, and was later scrapped in favor of a more sophisticated version.

The first American-made manipulators were adapted from the devices that are used in laboratories to handle radioactive materials remotely. This translation to the submersible was not an easy one, since the engineers involved were not familiar with sea environment and ignored a number of cardinal rules of marine design. These American manipulators, while quite complex in function and control, were not particularly successful. In several of these the error was made in using metals incompatible with seawater and other metals on the vehicles. One was damaged by serious electrolysis in less than one month and needed a complete overhaul.

One of the more successful early manipulators was built by General Electric for *Aluminaut*. This was the first use of a pair of arms that could be operated more or less like the arms of a human. This system has been highly successful in several salvage and recovery operations which will be discussed in the following chapters.

Among the problems encountered in manipulator design was that of the control system. Most manipulators have six degrees of freedom, the same as the motion of the human arm. This requires putting toggle switches on a panel or movable box, and the operator has to memorize a complicated control process. In some controls, microswitches and a joy stick were tried; in others, butterfly switches or hydraulic control were used. A simple control arrangement that could be learned quickly was found to be the best.

Some of the arms were electrically driven; others were electro/hydraulic or simply hydraulic. Whichever way the manipulator was operated was of little concern to the research scientist. He did care, however, whether the wrist would rotate continuously, since this was one way of making a rotary core of hard materials. Further, he wanted more than a simple "clam shell" or grabbing ability. The group of scientists at Woods Hole Oceanographic Institution developed a series of sampling tools that could all be operated as part of the basic 489-pound arm. These tools were interchangeable. A sample tray was placed beneath the arm in which bottles, corers, and such could be attached to the arm within the view of the pilot. *Alvin's* arm is capable of picking up 50 pounds at a reach of 63 inches. More complex manipulators were built for military missions like those of the *DSRV* and the Navy's submersibles, *Sea Cliff* and *Turtle*.

Each vehicle has a series of emergency devices and routines that it can use in the case of various accidents while submerged. Although differing slightly from boat to boat, they are basically all alike. Nearly all vehicles designed and certified within the past several years have an emergency breathing system, as described previously under Life Support. This system can either be built in, with masks readily available to each occupant, or the units may be separate mini-lungs. Although the apparatus may be portable, it is not intended for swimming out or for escape. The problems involved with free ascent—having enough gas to pressurize the inside of the vehicle and the small chances of successful escape—have precluded the design of such systems.

If one or more air-filled containers, such as instrument cases, should flood accidentally, causing negative buoyancy, there are usually several ways to drop ballast to gain more buoyancy. For all Navy-certified vehicles, it must be proven that the vehicle can recover and return to the surface in case of most accidents. Many vehicles can drop at least one battery, expel mercury, jettison air bottles, etc. By dropping these components, some vehicles may gain several thousand pounds of buoyancy. Perhaps the most effective system developed of this type has been the *Alvin* hull, which can be released from the rest of the submersible and is buoyant, forming a rescue capsule for its personnel.

# Submersible Components and Systems

This system has never had to be used under emergency conditions. Vehicles carrying special scientific equipment on "brows" have a release device of explosive bolts that allows complete jettison, while explosive cable cutters sever electrical cables. Several of these emergency buoyancy systems have been employed on dives where accidents have occurred. Descriptions of these dives follow in Chapter VII.

Several of the larger vehicles carry a small silver zinc emergency battery inside the cabin, in case the main electrical system fails, to handle emergency loads such as life support and communication. Those other boats that drop a battery for emergency buoyancy can usually continue emergency functions on the remaining battery long enough to reach the surface.

A few of the subs carry an emergency pinger which, if the electrical system fails or if other more serious accidents occur, will switch on automatically, giving off a high energy signal detectable over an 8-mile range on the telephone (UQC) communication frequency for up to one week. In this way, a rescue/recovery mission could find the downed sub quickly. For surface identification and location in fog and darkness, most vehicles now use the high intensity xenon flasher, visible for 2 to 5 miles. This unit is designed with a pressure switch that turns on the light at the surface.

While some boats use a special FM frequency radio to talk to their support ship, it has been found best to have an additional frequency on one of the Coast Guard distress bands that is guarded continuously in most countries. On one occasion, *Deepstar 4000* surfaced out of radio range of her mother ship during the night in a swift surface current and was unable to converse with other ships nearby for assistance. A radio with a Coast Guard frequency would have helped. In the case of *Deepstar,* the pilot had to use a flare gun. Most submersibles carry an emergency kit consisting of a flare gun, a life raft, and rations. None of the existing vehicles has yet to require the use of this life

115

raft kit, but both *Alvin* and *Deepstar* have had an emergency situation during which several of the other emergency systems were used.

Probably the most important part of a total submersible vehicle system is the vehicle support/handling system. It is inescapable that any kind of submersible launching and retrieval made at the air-sea interface is bound to be hazardous except in a calm sea. Captain Cousteau stressed this fact in 1965 when he declared, "Avoid the interface between sea and air. Either work above or below, but not in it." Nevertheless, nearly all support ships and handling systems developed in the 1960's were relatively conventional—all working in the interface of sea and air where conditions were worst.

One of the better systems was developed by Westinghouse, using an offshore oil-well service ship with the largest available earth-moving backhoe modified for a lifting crane. This 165-foot ship has room for work and living quarters in trailer vans, and yet is one of a large class of similar ships relatively inexpensive to lease and operate. From this type of platform, both the *Soucoupe* and *Deepstar 4000* made over 650 dives, and *DOWB* and *Beaver IV* also made a few dives.

The ship and crane permitted making dives in conditions up to a sea state 4, that is, wave heights of 4 to 6 feet and wave periods of over 7 seconds. The heart of this system was the hydraulically controlled crane, capable of lifting about 20 tons from the water. Divers were used to place a large hook in a lifting eye fastened to the vehicle. At this point, the ship facing stern into the seas might be moving at a different motion, due to the action of the seas, than *Deepstar* which, nearly submerged, moved more at the frequency of the waves. The crane operator would pick up *Deepstar* at the moment when wave and ship motion were most closely matched, and would then rapidly lift the vehicle out.

Although no accidents ever occurred with this system, there

were times of great stress on the crane and vehicle, as well as the pilots and crane operators. On one occasion, an automatic release hook holding *Deepstar* with pilots aboard released prematurely, dropping the boat 4 feet into the water—without serious harm. This support system, although successful and maybe the best to date, was not ideal. Further, it proved to be costly over the total period of use. Perhaps more than half the cost of the whole submersible system is for the support ship and its handling gear.

Other systems specially designed to overcome some of the above shortcomings have not been much better. *Lulu* is the support catamaran for the *Alvin*. It was conceived and designed by Woods Hole and built on a shoestring budget by a local contractor using surplus 96-foot pontoon hulls joined together. *Alvin* was handled from the center on a platform raised by four cables. The ship was self-propelled at a slow 4 to 5 knots, and was not highly satisfactory in rough head seas. However, *Lulu* has turned out to be reasonably good after several years of modifications. Unfortunately, it was this system that was responsible for losing *Alvin* overboard when a cable failed in 1968. The *Lulu* catamaran has operated in heavy seas of sea state 4 or greater, but with some danger to vehicle and personnel.

Two other newer launching techniques are worth considering. One is the *Midwife* concept of Westinghouse, built in 1970 for *Deepstar 2000*. Instead of using a large, fairly expensive ship and crew, this catamaran is 45 feet in length, self-powered, and can be disassembled, for road transport. The ship is so constructed that it is hydrodynamically matched or "tuned" to the surface motion of *Deepstar*, making launch and retrieval easier. While there are no living quarters aboard for extended diving periods, a second small vessel would accompany *Midwife* on overnight trips. This system could prove efficient and economic for local diving.

A more advanced idea that many submersible pilots have had

117

in mind for years is a submersible launching platform. One was built in 1970 for the Makai Range in Hawaii. The Launch Retrieval Transport (LRT) is a catamaran hull with a center platform about 27 feet long. The submersible to be launched is fastened down on the center deck and the LRT is towed to

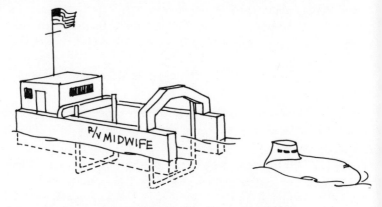

**Support Ship** *Midwife* **for** *Deepstar 2000*

the launch site by a small tow boat. Here, after preparatory checks the entire LRT is submerged by taking on water ballast. The sub is released by the divers at a depth of 50 feet or so, where the wave action is greatly diminished. Thus, even in rough seas where surface conditions would prohibit launch and retrieval, the LRT makes it possible to make submerged launchings. For retrieval the LRT meets the submersible at the 50-foot depth and the process used at launch is reversed. A larger version of the LRT is being designed for the U.S. Navy to be used for *Turtle* and *Sea Cliff*.

Surface support for the larger submersibles has been more of a problem. *Aluminaut* is 49 feet long and weighs nearly 80 tons. None of the available mother ships could lift it from the water so, instead, it is towed to its dive sites at a speed of 6 to 8 knots, depending on the conditions. In rough seas, the speed may be as slow as 4 knots. On one occasion, *Aluminaut* was loaded on a

# Submersible Components and Systems

136-foot minesweeper and transported from Florida to New England where it was unloaded by crane at a port, and then towed to the dive area. Surface towing limits speed and makes for higher costs. Yet, in the case of very large vehicles, it is the only solution short of building a specially designed support craft.

The previous components, when taken together in an assembled total system, make up the working modern submersible. Now that we have seen all the parts, let's look at the whole submersible and what sort of work there is for it.

# 6

# Work for the Manned Submersible

The advent of the manned submersible was propitious for oceanographic research and for other underwater work waiting to be done. The greatest advantage gained by putting a man underwater in a mobile vehicle was that he could be face to face with the environment he had rarely if ever been able to observe before.

Professor William Beebe's reports on his remarkable observations from Otis Barton's bathysphere in 1932 and 1934 were the first accounts of what could be seen with the human eye at a depth of 3,028 feet. He demonstrated the value of visual observations of biologic organisms and their behavior as seen through a port while he dangled on the end of a long cable in mid-water.

With the availability of free-moving vehicles in the decade of the sixties it was agreed by scientists that visual capability

121

was the prime advantage for nearly every mission. During the demonstration of new vehicles being offered by major industrial firms, one dive was enough to enlist the enthusiasm of a scientist for he could see how he could use a deep research vessel (DRV) for direct observation in his field research which previously had to be conducted from a surface vessel.

Until the appearance of the DRV, scientists had relied heavily on indirect measurement and interpretations of conditions which they presumed to exist on the bottom or mid-water. Of course, many of those theories may have been partly right, but there was the feeling of dissatisfaction in not being able to see for oneself. To overcome some of this inadequacy, some extremely good cameras were developed to photograph bottom features. Also, television had been perfected and could be used in shallow depths. Yet neither of these techniques was able to approximate what a man placed in the same spot could see and do after making a decision. The ability to look about, scan freely, and maneuver around an object was not possible with TV or film cameras. Several attempts with remotely operated unmanned vehicles carrying TV and still cameras showed the severe limitations in viewing and in vehicle control. Few persons who have been in a submersible with adequate ports would fail to agree that good vision is the key to success in undersea research.

A second advantage demonstrated by the manned submersible is the ability to collect samples and be selective about it. Previously a geologist could take a grab sample or a sediment core with no difficulty, but he couldn't know its characteristics until it was returned to the surface. He really could not tell the exact location on the bottom nor identify the environment from which he took his sample. The geologist trained on land is accustomed to describing the surrounding terrain in his field notebook, and at last an underwater geologist could do the same. Most of the submersibles used for oceanography have some sort of mechanical arm or manipulator which let the scientist reach outside his protective cabin and pick up a hand specimen. This sample could later supplement his visual observations. For the first time, the scientist, who had had a passive role, could take an active part in his research.

A third distinct advantage of the use of a submersible is in the installation and monitoring of oceanographic instruments. Once a vehicle demonstrated that it could maneuver into small areas, carry instruments, and perform tasks in the ocean environment, a whole area of specialized measurements was opened. Some equipment, like a current meter, requires care in installation so that precise, reliable results are assured. The DRV can be used to install an instrument and make sure it is placed where the scientist wants to make measurements—for example, a current meter behind a boulder in a canyon would not give a correct record of water movement. Once installed, the vehicle can return, check appearance, fouling, or nonperformance. In some cases, it is possible that a repair or replacement can be made by using the submersible's manipulator. Such a task would save considerable money compared with retrieval from the surface and reinstallation. Furthermore, in performing this kind of work the vehicle serves as a test platform for new instrumentation, a use that probably is uneconomical with the more expensive subs.

Although it might be possible to make long lists of the advantages of manned submersibles, one final one should be sufficient to make the point clear. An undersea vehicle can carry a scientific payload greater than is possible with surface wires, at the same time allowing a fully integrated set of observations using photographic, visual, acoustic, sampling, and measurement techniques. A vehicle far beneath the sea's waves is far more stable than a ship on the surface. Finally, it has been calculated that a deep (20,000 foot) survey and search vehicle can make greater speeds over the bottom than is possible with a towed device from the surface.

It is clear that many advantages have been realized from using manned vehicles. This is not to say there are no disadvantages; there are several as we shall see shortly. First, let's examine some of the types of diving missions in detail and some of the specific uses for the nearly seventy existing submersibles.

The scientific dive mission has accounted for a large percentage of all submersible use. This is natural since the first use of the diving vehicles was by ocean scientists who saw them as a way to extend their exploratory activities. Among these men were the biologists, who during the mid-nineteenth century were the first scientists to explore the oceans as they donned copper helmets to observe animals *in situ* in shallow waters.

The very simplest of submersibles was adequate for many biologists who were ecstatic at having a window from which to look out on the organisms in their natural habitat. The DRV allowed an investigator, with nose pressed to view port, to observe a particular species in detail, determine its distribution and behavior as well as the bottom habitat of the organism. Biologists wished to photograph animals and estimate population density and community characteristics. Neither cameras nor nets operated from the surface had been able to do this adequately.

One of the areas that had long perplexed the biologist was the species composition of the well-known deep scattering layer (DSL). This generally shallow series of layers had been known for twenty years from acoustic reflections of ships' sonar—the sound energy bounced off layers or concentrations of various small fish, squid, and jellyfish. Attempts to photograph in these layers of the upper 500 meters of water were disappointing. The ability of a submersible to hover motionless in a particular layer while a biologist looked at the individual animals that cause the sonic scattering was important. *Trieste I* attempted to make observations of these layers on her dives to great depths. Frustrated biologists like Eric Barham of the Naval Underwater Center, hoping to see some signs of the wee beasties, saw nothing but swirls of water as the monstrous hulk plummeted through these upper layers like an express freight train through a sleeping village. He saw none of the animals he knew must be there. A slower descending vehicle like *Soucoupe* or *Deepstar* was more suited for this purpose.

Other biologic missions have involved looking for whales, porpoises, or dolphins in their natural habitat. The submersible has not been particularly adept at this sort of task since it is unable to pursue any quarry at more than several knots.

Beyond the desire to observe visually the activities of marine organisms, many biologists have been eager to collect what they saw through the ports. Dick Rosenblatt of Scripps Institution of Oceanography, after several dives in *Soucoupe* at Cabo San Lucas, Mexico, commented, "I got so excited looking at those fish I had a desperate urge to reach right out to grab them—and more of a frustration knowing I couldn't."

Later, devices were built and rigged on several submersibles that resembled giant "slurp" guns or vacuum cleaners that could suck in small fish for the scientist to collect. This was not a perfect solution but it helped to collect several interesting specimens not seen before. The submersible allowed a scientist to sneak up on a whole community and collect species that had up until then eluded nets by hiding in crevasses and behind rocks.

One of the most significant developments to the biologists is the lock-out hatch submersible that allows a diver to swim out breathing a gas mixture supplied by hose from the vehicle. A trained biologist can collect valuable samples by hand instead of using a manipulator. This method is limited to depths less than 1,000 feet and has been used to 750 feet so far.

Biologic dives in fishery research work have used DRVs to make observations. Some of the earliest work has been done by the Japanese and Russians, both having large aggressive national fisheries programs. The Japanese tethered bell *Kuroshio I* was used from 1951 to 1960 to aid the fisheries in gathering information on bottom conditions favorable for fisheries. The Russians used a converted fleet submarine, *Severyanka,* to make numerous dives for observation of fish behavior and measuring physical properties of the water. They also saw how fish avoid towed nets.

The biologists who have used submersibles over the last ten years found that the vehicle cameras were about the most important feature. With still and movie cameras, it is possible to document and record all the observations. This, together with a field notebook and tape recorder, represents 80 or 90 percent of the dive results.

The pursuit of geologic investigations by submersibles is similar to the biologic. Both rely on direct observation and selective sampling. The geologist on land is accustomed to walking along the outcrop of a bed or strata to map an area. His first chance to do this underwater came with the introduction of scuba in 1950. But scuba was much too limiting in depth, duration, and range. The submersible allowed the marine geologist to go far deeper in "walking the outcrop," and has turned out to be better suited to the geologist than to any of the other scientific disciplines. In the United States twice as many geological dives were made as biological dives, whereas in France the opposite was true. The Russians were the first to make a geologic dive when, in 1926, Mme. M. V. Klenova, a noted marine geologist, dived to 150 feet to collect samples.

The first deep geological observations reported from a submersible were by a group of American geologists led by Dr. Robert Dietz, then with the Office of Naval Research. These men had a chance to dive in *Trieste I* in 1957 off Italy. Their enthusiasm for what could be done in observing features on the sea bottom convinced the U.S. Navy to purchase *Trieste I* and to begin a serious program of sea floor studies with submersibles.

Geologists were able to dive to great depths in the bathyscaphe *Archimede* to explore trenches off Puerto Rico. But they found that bathyscaphes were unsuited for maneuvering in close quarters or for long bottom transits. *Trieste I* did however open the

eyes of scientists to other smaller submersibles. Geologists were
able to use this bathyscaphe to follow the course of submarine
canyons far offshore. The major emphasis was to study the steep
slopes of crustal blocks such as the Coronado Bank, where out-
crops of shale and interbedded sandstone were seen.

Later vehicles such as *Soucoupe, Alvin,* and *Deepstar* were
small enough to explore along the gullies, canyons, and steeper
continental slopes where the geologist was interested in progress-
ing slowly down the shore slope, to look for marine terraces that
he had seen indicated on echo sounder traces made from the surface. These terraces were benches made by a lowered sea level
which remained long enough for the sea action to cut a nip or
terrace during earlier geologic time. While it was possible to spot
some of these from a surface ship, there had been no way to inspect, photograph, or sample them to confirm that these were
marine terraces. By properly identifying the terraces and their
exact depths, geologists could reconstruct the histories of islands,
continents, and gain information on how sea level has varied on a
world-wide basis.

Perhaps the area where the submersible has played the most
important role in marine geology is in the study of submarine
canyons and their origin. The *Soucoupe* turned out to be ideally
suited to give marine geologists such as Dr. Francis Shepard of
Scripps and Dr. Robert Dill of Navy Underwater Center the opportunity to study the narrow valleys, overhanging cliffs, and
sheer walls found in many canyons along the California and Mex-
ican coasts. Until the arrival of the *Soucoupe* in 1964, geologists
equipped for scuba diving had explored only the heads of canyons such as Scripps Canyon, La Jolla, and Cabo San Lucas.
Using the *Soucoupe,* Shepard found that overhanging walls and
narrow ravines continued for the entire length of the canyon.
Scripps Canyon became so narrow at a depth of 650 feet that the
10-foot-wide *Soucoupe* could not squeeze between the walls at
the bottom. Dr. Shepard said he had seen more in one of his

four-hour dives in the *Soucoupe* to help establish his theories on submarine canyon formation, than he had gained in nearly thirty years of probing these canyons with dredges and cameras from the surface. The submersible allowed a geologist four hours of observation time in which he could carefully inspect the canyon walls and floor for abrasion marks, collect sediments or rock fragments, and abundantly photograph all he saw.

While the submersible answered the need of the canyon explorers, it also interested the geologists who wanted to sample, core, and investigate the deeper slopes. One of the problems geologists faced in merely observing was that much of the bottom they traversed was covered with an overburden of silt, clay, or sand. In some places they were able to poke through a thin dusting of sediments to see and perhaps obtain a sample of the underlying rock. In other places where the cover was encrusting growth or semihard sediment, a drill or rotary corer was needed to penetrate this layer. In some cases, a subbottom profiler could penetrate the overburden showing the nature of the bedrock beneath and where it might be best to try to sample. Those submersible-operating groups who have kept their vehicles equipped with the geologist's special tools have also managed to keep their vehicles busy and productive. The geologist had advanced quickly from being content to simply look out through a port and take some photographs to demanding the ability to gather a reasonable sample collection. Some geologists are not even content to collect samples to be examined upon return. They prefer inspecting the nature of the sediment *in situ* in order to know exactly what it is they are collecting and also to prevent changes from occurring in the sample when it is removed from its environment.

The geologist continues to require good maneuverability and accuracy in positioning. While a number of the presently available vehicles can meet maneuvering requirements for working in canyons and trenches, few can determine their location with suffi-

The Westinghouse/Cousteau *Deepstar 4000* during scientific dives for the Naval Underseas Center, San Diego. *Deepstar* is a three-man vehicle, weighing 18,000 lbs., and has been versatile in diving missions in the Pacific, Gulf of Mexico, and the Atlantic for over four years on 540 dives. (Joe Thompson)

*Deepstar 4000* with fiberglass fairings removed for servicing. Note the blocks of syntactic foam cut to fit and fill the voids. The large 900-pound forward battery case can be jettisoned for emergency buoyancy. (Joe Thompson)

The interior of *Deepstar 4000*. The pilot, left, and observer, right, lie on adjustable couches, each looking out a port. Movie camera between them uses a third port. Scientific equipment is rack mounted to the right and above observer. Pilot controls are all at close reach while pilot views exterior directly. (Joe Thompson)

*Deepstar 2000* was designed and built by Westinghouse Electric Corporation based on the principles of *Deepstar 4000* and other Cousteau vehicles. She is capable of diving to 2,000 feet with three men for scientific exploration and oceanographic missions. Her initial sea trials and about fifty dives were made in San Diego but she is now based in Annapolis, Md. (Joe Thompson)

Several submersibles have been placed aboard the offshore service ships similar to the 136-foot *Burch Tide,* for extended mobile operations. The diving vehicle is the main mission and all equipment supports it. Here seven trailer vans, a towed sonar, and a specially modified backhoe crane are part of the diving operation. (Westinghouse Electric Corp.)

Many of the smaller submersibles like *Pisces III* can be transported by commercial aircraft. Large subs like *Deep Quest* or *DSRV* can be airlifted by military aircraft. (HYCO)

One of the smallest of all manned submersibles is *VAST Mark III*, a one-man vehicle designed for 250-feet depth. Here the *VAST* is diving in the Caribbean for the U.S. Navy.    (Perk Bingham)

*PC-3X Sub Rosa*, the first Perry *Cubmarine*, being used here in the Tongue of the Ocean, Bahamas, August, 1962. This 150-foot-depth vehicle was rebuilt in 1969 after 500 dives. She was renamed *Gaspergou* and now is used by the University of Texas.    (E. W. Hull)

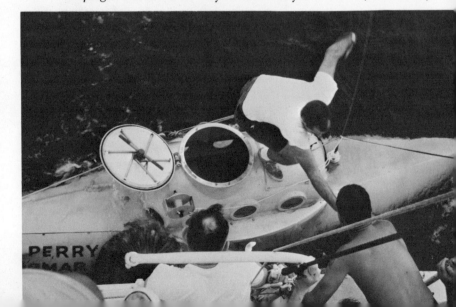

The Pisces family of International Hydrodynamics, Inc. (HYCO) Vancouver, B.C., lives up dockside. *Pisces I* and *III* are operated by HYCO, while *Pisces II* is owned and operated by Vickers of England. *Pisces III* dives to 3,500 feet and has made 325 dives. *Pices IV*, a recently constructed 6,600-foot-depth vehicle, was purchased by the USSR Institute of Oceanography.   (HYCO)

One of the two Cousteau Mini Subs. These are one-man vehicles that dive to 2,000 feet and have been used for exploration and filming over the last two years as part of Cousteau's television series. (Joe Thompson)

Lockheed's *Deep Quest,* an 8,000-foot-depth vehicle, and one of the most complex and sophisticated of the submersible fleet, has made over 130 dives since launching in 1967. *Deep Quest,* 40 feet long, weighing 55 tons, is supported by the 105-foot *Transquest.* (Lockheed Missiles & Space Corp.)

*Alvin* has had one of the most adventurous careers of any of the submersibles. *Alvin* was most recently salvaged from 5,000 feet after accidentally plunging from her support ship *Lulu.* She is operated by Woods Hole Oceanographic Institution and has made over 320 dives for a variety of scientific missions. (WHOI)

*Trieste II* bears little resemblence to the first *Trieste* that dived to 35,-800 feet in 1960. She is operated by the U.S. Navy Submarine Development Group One, San Diego. *Trieste II* has made over 75 dives in this present modification, one of these to 13,100 feet.   (U.S. Navy)

U.S. Navy's *Deep Submergence Rescue Vehicle* (*DSRV*) was designed to mate to a downed combatant submarine and rescue crew members. It can carry twenty-four rescuees per trip. *DSRV* is the only submersible that can make an underwater transfer of persons. *DSRV-1* and *2* are both stationed in San Diego, California and operated by the Submarine Development Group One.   (U.S. Navy)

This view, the pilot's seat of the cockpit of the *DSRV-1*, shows the complexity of instrumentation and control in contrast with the smaller, far simpler submersibles. It takes a crew of three to operate the *DSRV*. The vehicle is normally controlled by a computer but the two ball-shaped manual controllers can be used to control vehicle in six directions of motion. (U.S. Navy)

One of the Navy's two *Alvin*-type submersibles, *Sea Cliff* is a 6,500-foot-depth vehicle now operated by Submarine Development Group One, San Diego. This three-man sub has two manipulators, a trainable stern propeller, and rotatable side-pod propeller. *Sea Cliff* and sister, *Turtle*, were built by General Dynamics/Electric Boat Division. (U.S. Navy)

*Turtle* is slightly larger than *Alvin* although bearing close resemblance. She weighs 48,000 lbs. This submersible, like sister ship *Sea Cliff*, is operated by the Navy's Submarine Development Group One in San Diego. Note · the two manipulators and tool rack. Both vehicles are diving on Navy projects in the Pacific.   (U.S. Navy)

*AGSS-555 Dolphin* was launched in 1968 and is operated by the U.S. Navy Submarine Development Group One, San Diego, for oceanographic and sonar research. It can dive to great depths and carries twenty-two men and five scientists.   (U.S. Navy)

*Ben Franklin,* the Grumman Aerospace vehicle built by Jacques Piccard in Switzerland for drifting in mid-water. Here in Palm Beach, Florida, *Ben Franklin* was being prepared for a drift mission in the Gulf Stream in 1969.   (Grumman Aerospace Corp.)

*Ben Franklin* undergoing trials in Florida prior to its 1,443-mile drift mission where five men spent 31 days submerged at about 2,000 feet, drifting northward in the Gulf Stream.   (Grumman Aerospace Corp.)

One of the compensated lead acid battery cases used on *Deepstar 4000* showing the individual cells immersed in oil. The valve at the top allows gas discharged by battery cells to pass freely to the seawater but prohibits water from entering the battery. (Joe Thompson)

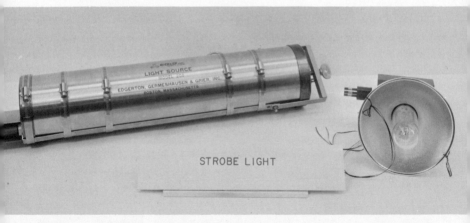

STROBE LIGHT

The strobe light is necessary to give adequate lighting for underwater photography. The strobe unit with reflector is placed to the side of the camera and at angle and prevents a bounce-back from scattering particles. The electronics unit can be mounted externally in a convenient place. (EG & G)

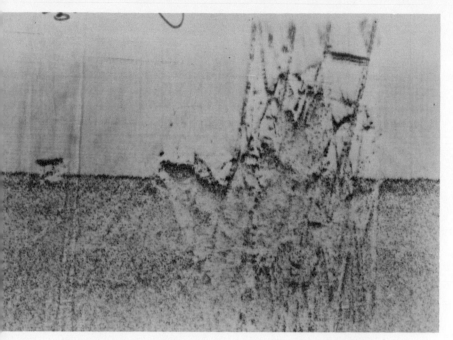

Larger submersibles have a side-looking sonar used for searching the bottom. It presents an acoustic picture of the outside in plan view. Here is a record from a side-looking sonar showing an oil rig that blew over in a hurricane in the Gulf of Mexico. (Westinghouse Electric Corp.)

Speed sensor using a Savonius rotor is mounted on the *Diving Saucer* to measure vehicle speed when moving at current speed while stopped on the bottom. A temperature sensor was placed on top of the speedometer. (Joe Thompson)

A forward scan sonar developed for smaller vehicles. This sonar is a puke type and uses less power and is lighter than the CTFM type. The display is a plan view showing objects up to 500 yards ahead. The scanning transducer (right) is mounted externally. (Westinghouse Electric Corp.)

One of a variety of 35mm and 70mm cameras used for underwater photography. Most cameras take up to 400 exposures. Some have zoom and focus adjustments. All are mounted externally and coupled with a strobe flash for lighting. (EG & G)

The underwater telephone has become an essential part of the instrumentation in a submersible. This one is an example of a high-powered unit using the UQC frequency and latest communication techniques. (Westinghouse Electric Corp.)

*Beaver IV*, also called *Rough Neck*, was North American Rockwell's vehicle designed to work in oil production activities. *Beaver* can dive to 2,000 feet, can lock-out swimmers, and perform complex tasks. She has made 72 dives and is presently in storage due to lack of money for further development, or government use. (Joe Thompson)

High visibility is possible with this acrylic hemisphere. This one is being used for the Perry *PC-8*, a 600-foot-depth vehicle built by Perry Submarine Builders in Riviera Beach, Florida. The acrylic window offers 120° forward view with minimal distortion. A similar acrylic plastic window is used by the *Johnson - Sea - Link*. (Perry Submarine Builders)

A new vehicle experimenting with glass hemispheres for greater viewing is *Deepview*, built by the Naval Underwater Research and Development Center, San Diego, California. *Deepview* will dive to 1,500 feet. The forward hemisphere is cast glass, held in place by a titanium framework. *Deepview* began sea trials in late 1971. (Joe Thompson)

cient accuracy required for surveying and mapping.

The physical oceanographer is primarily concerned with the physical properties and water motion of the ocean on a large scale, and has some interest in small-scale effects within the water column and the near bottom.

One of these concerns is current measurement. Current measurements can be made by an instrumented vehicle, and generally a current record of several weeks is required for a proper evaluation. The vehicle may be used to implant a meter and periodically observe and service it. In places where the circulation is too slow for a meter, dye has been used—but only for short periods of investigation. One way to carry out a long-term current measurement is to make the vehicle itself a neutrally buoyant float that can drift with the whole mass of water in a major current stream. This sort of experiment was carried out with the *Ben Franklin* in 1969 in the Gulf Stream Drift Experiment.

Several submersibles have been equipped to measure the standard variables of temperature, salinity, and pressure. *Alvin* was equipped with an STD (Salinity-Temperature-Depth) instrument used by Woods Hole scientists. For the most part, this application has been unsuccessful compared with the accuracy, speed, and cost of surface techniques.

However, using similar techniques, microchemical and thermal properties have been measured near the bottom and at the bottom sediment-water interface, and in the sediment to determine how these factors affect the transmission of sound in bottom water. Such information has been of great interest to the U.S. Navy's programs in acoustics.

The study of ambient sea noise has been helped immensely by the use of a submersible. Some of the advantages of measuring the background or ambient noise in the ocean by a vehicle *in situ* include the elimination of the surrounding power plant noises of a ship as well as the pitch and roll which affect direction determinations, not to mention the noise of waves lapping against the ship's hull. A number of successful noise measurements have been made by submersibles over the last ten years and these proved to be less expensive and more accurate than measurements made from the surface.

The study of sound in the ocean, especially near the bottom, is of much interest to the basic research necessary to the pursuit of anti-submarine warfare (ASW). While the end result is military in nature, the study of the behavior of bottom acoustics is part of continuing scientific studies, and so the submersible has been used to study the way sound is reflected by different bottom materials. This type of mission requires a vehicle which can carry a sound source to be projected at the bottom. The bounced signal is then measured by the submersible's surface tender.

Some deep submersibles have been used for target strength measurements in which the DRV simulated a military submarine to be detected by a surface ship sonar. In most cases, the DRV moved too slowly and was too small to act as a real target. One of the largest experimental vehicles of the Navy is the *Dolphin* AGSS 555, a 150-foot-long submarine which will be better able to carry out such maneuvers for sonar development.

Another property that has been of interest to the physical oceanographer is visibility and light penetration. The submersible is a stable platform for making these measurements, whereas the surface ship is subject to roll and drift. Further, the submersible can conduct horizontal visibility studies in various layers of water as well as water visibility beneath the vehicle, both of which would be difficult by remote instrumentation.

Associated with visibility is the amount of suspended sediment and other particles. Suspended sediment is of interest to oceanographers because it controls the optical properties. Different layers are often identified by the amount of suspended matter. In some areas, the intermediate layer is usually dense, whereas in other areas, the bottom layer has poor visibility due to plankton or sediment. From a manned vehicle, it is possible to note these changes and make measurements and take samples.

The geophysicist is primarily interested in measuring the earth's gravity and magnetic fields to help him determine major earth structure. A specific and practical application of geophysics has been demonstrated in petroleum and mining exploration. As early as 1923, Vening Meinsz, a Dutch geophysicist, used a pendulum gravimeter to conduct a survey from a submarine. The submarine hovered or moved slowly beneath the wave turbulence to provide a stable platform for the instrument. When the DRV came along, it was natural to experiment with a gravimeter. Although these instruments have been tried aboard most of the smaller and larger vehicles, it was found that the instabilities and sudden accelerations of smaller boats such as *Alvin* and *Deepstar,* as well as their very limited horizontal range, made them unsuitable for this sort of work. The larger boats, such as *Aluminaut, Auguste Piccard,* and *Ben Franklin* are better suited for making some worthwhile gravity measurements that can be put to use in local gravity mapping. The U.S. Navy submarine *Triton* made periodic gravity measurements during her round-the-world underwater cruise. A number of other military subs have been used for shallow water gravity observations.

On the other hand to make magnetic measurements from a submersible is to put it to poor use, since magnetic observations are not affected by ship motion and are far less expensive to make from the surface.

A geophysical mission requires a large vehicle with large power supplies capable of accurate positioning and of traveling considerable distances, in short more than any of our present scientific submersibles have in capability.

While probably 80 percent of all the diving with DRVs has been done for the scientific missions just described, some of the most promising missions have been engineering and salvage dives.

131

Submersibles that can search a limited area aiding in salvage have been of importance in several military operations and civilian salvages. The submersible has proven particularly suited for working in deep water and performing complex functions with its manipulators such as attaching salvage lines or hunting for small objects. The first of these operations was the search for the downed submarine SSN *Thresher* in 1963, followed in 1966 by the search and recovery of the H-bomb in Spain, and in 1969 in the search and salvage of *Alvin* by *Aluminaut*. The larger vehicles such as *Deep Quest* have been designed to carry out search missions far better than the smaller *Alvin* or *Deepstar* types. Search requires the ability to cruise long distances with relatively good navigational positioning, enough power for extended cruising and for using sonar, TV and video tape, and other sophisticated equipment. Several searches for aircraft wreckage have been efficiently carried out by *Deep Quest,* which showed the value of a manned submersible in this type of mission, as a tool complementing other surface-towed search devices.

Another nonscientific use was for making engineering site surveys. The vehicle with special instruments can make necessary soil strength measurements prior to placing large installations. These surveys determine sediment-bearing strength, sheer strength, and soil stability in a localized area, just as any engineering site survey does on land by taking borings before beginning to build. The ability of the vehicle has been demonstrated but as yet there has not been adequate development of measuring tools to make this type of mission practical. Much of the development of tools such as vane-shear devices and free-fall cones to measure sediment strength have been experimental and are not yet standardized.

An area of great potential for the submersible lies in its use in the petroleum business, yet to date few vehicles have proven their value to this billion dollar industry. Several of the small- and medium-sized subs have been demonstrated and leased to oil companies for experimental exploration work, salvage use, and certain production tasks. The only venture which was specifically designed for work around a well head was the unsuccessful attempt by North American Rockwell to build *Beaver IV* capable of performing for Mobil Oil Corporation. The most simple explanation why the oil companies have hesitated to use submersibles is that it is a matter of economics, and existing surface and dive techniques are still quite competitive.

Related closely to the petroleum industry is the need to survey, inspect, and repair pipelines carrying gas and oil. The recent purchase of a 1,200-foot-depth Perry *PC-9B* by Brown and Root Construction is an indication of a serious interest in this field.

The U.S. Navy has been one of the chief forces in the development of the small manned submersible in this country. It is difficult to distinguish between the Navy's scientific diving programs, already discussed, and the purely military missions.

The main thrust of the Navy program has been one of personnel rescue and salvage of deep objects carried out by the Deep Submergence Systems Program. In addition other missions for military oceanographic survey and ASW acoustical work have formed the requirements for vehicles considerably larger than the DRVs yet smaller than conventional combatant subs. These requirements are for seven to twenty-five man crews on cruises of several days to a month—and for the most part fall into a submarine classification where there is no dependence on surface support. Most of these submersibles developed during the 1960s are just becoming operational and more to be considered vehicles of the future. Each will be described in Chapter Nine.

From the previous discussion, it may seem that the research submersible is the answer to all the oceanographic problems and these vehicles ought to be plying our oceans continuously since they offer so many advantages over surface ships. Let's consider the factors that limit their operation and some of the requirements to make a complete submersible system effective. Some of these factors include logistic and ship support, seaworthiness, endurance, depth, reliability, and cost.

As anyone who has worked with operations at sea knows, success depends on a logistics system sufficient to keep everything supplied and functioning. In the case of a submersible, the support system is often as important as the vehicle itself. Two basic approaches have evolved. One group of vehicle operators believe that the submersible is an extension of the research ship. Since the ship costs more than the vehicle to maintain and operate, the ship is operated at full potential; that is, it is instrumented to carry out large-scale surveys, work around the clock, and do other research besides operating the submersible. The ship in this system is equipped with a crane for the vehicle and all the vehicle-support facilities, as well as standard oceanographic winches, sub-bottom profilers, cores, samplers and oceanographic laboratories. While this system has often been proposed, it has not been adopted by many research institutions. Scripps Institution of Oceanography is planning to use *Star III*, donated to them in 1970 by The Electric Boat Division of General Dynamics, in this manner.

The second approach could be called the vehicle-oriented support system. The submersible becomes the central element in this system with everything on the ship designed around the vehicle. The mother ship either carries the vehicle or tows it. If a mother ship has been specially designed to launch and handle a particular vehicle, it means that the vehicle is limited to operating from that ship alone. An example is *Transquest*, the 105-foot specially designed catamaran support ship for 40-foot *Deep Quest*. It is unlikely that *Deep Quest* could be operated from any other ship because of its size and weight. If there were a requirement for that submersible a long distance away, it would have to be taken by sea at 5 knots, which is time consuming and costly.

134

# Work for the Manned Submersible

An alternate approach used by several operating groups has been to lease a commonly found class of ship sometimes called boat trucks or offshore supply vessels used around the world in the petroleum industry. This class of ship is built in sizes up to 165 feet and is capable of crossing most oceans. Westinghouse Electric Corporation has made extensive use of these ships in support of their *Deepstar* program. A crane is mounted on the stern and the open deck space houses living and maintenance vans. While this system is more versatile than the specially-designed support ship, it has not been practical to ship by air a whole vehicle support system including crane and vans. However these ships are quite capable of long sea voyages at 12 knots— Westinghouse took *Deepstar* from San Diego to Nova Scotia and back by sea.

A further limit in working with conventional ships that must lift vehicles out of the water is the mismatch of motions of ship and submersible in the sea. The sub which is usually 90 percent or more submerged, moves at a different rate of motion than the ship which may have less than half her mass exposed to wave action. If facing stern to sea the ship also tends to pitch at the moment of lifting the sub from the water. In short, in choppy seas of over 4-foot-wave height, severe accelerations develop in the lifting system. One way to overcome this is to partially submerge the support ship to dampen its motion and thus dynamically tune the two vehicle motions closer together. Another solution which theoretically is far better is to avoid the surface interface altogether and operate beneath the sea.

A still better solution for submersible operations is to eliminate the surface ship completely and return to a submarine that can travel at will, submerge, and be fully independent of surface support. While this may be ideal from the support angle, it requires a larger submarine at a greater investment and precludes the very advantages that make a small, maneuverable boat attractive to the scientist.

135

A final part of the logistics system that should be mentioned is the need for repair and service facilities. When Westinghouse was operating the *Diving Saucer* in Mexico in 1965, it was found that although apparently well supplied with spare parts and a complete maintenance shop aboard the 136-foot support ship, there still weren't enough spares. A charter plane had to fly back to civilization to have a motor armature rewound that had failed for the third time. When a submersible is going to be operated reliably away from home base, a large stock of spares is mandatory.

The submersible which must be towed to the dive site because of its large size or because of the lack of a handling system must be designed seaworthy. The basic shape of vehicles such as *Alvin, Deepstar,* and *DOWB* simply precludes practical towing.

In general the long, narrow hulls can be towed easiest. For most situations towing has a limiting effect and is used for near-shore short range operations where a proper handling system is not available. Several of the smaller boats are towable at 8 to 10 knots under low sea conditions. Some of these are the Perry *Cubmarine. Aluminaut* has been towed from Florida to Venezuela and back at a moderate speed.

The endurance of both man and machine is also a limiting factor. In the smaller boats, there is no way for crew rotation and rest. Usually, these vehicles also are short on electrical power and must return to the surface after four to eight hours. Most of the scientists using *Deepstar, Alvin,* or *Diving Saucer* agreed that after intensive observation, note scribbling, talking into a tape recorder, and conversing with the pilot over a high noise level, four to six hours was ample for one day. Other missions, such as watching the occurrence of the deep scattering layer (DSL) may take much longer periods, perhaps up to twenty-four hours. For this kind of work the larger boats such as *Deep Quest, Aluminaut,* and *Ben Franklin* are more suited for longer missions. The *Ben Franklin* was specifically designed for long-endurance missions. She was built for Grumman Aircraft in Switzerland under the direction of Jacques Piccard. The design goal was a vessel that could stay submerged for periods of from forty-five to sixty days. The *Franklin* is designed to use very low levels of power while on a thirty-day dive. The endurance of the submersibles is purely dependent on the life of the power sources such as batteries.

All vehicle designs involve a series of compromises of the best-suited alternatives for the intended purpose. Speed is one factor that has given way to obtain greater endurance. Most of the scientists in the first-generation submersibles were content with speeds of 0.5 knots to 2 knots. When inspecting the bottom, half a knot (0.6 miles an hour) seems fast enough. Any greater speed and the observer might begin to miss objects of interest. Investigators and ocean engineers involved with newer vehicles have demanded higher speeds. It turns out the amount of power required for small speed increases is enormous. In general, double the speed of a submersible requires *eight* times the output from the propulsion system. This means an enormous increase in battery supply which in turn increases the weight and size of the vehicle. Suddenly the designer finds his boat has gained too much weight. Some solutions used have been to reduce the drag by streamlining and using an elongated shape. Therefore, speed and endurance are probably the largest limiting factors in submersible design because each demands more power and thus more weight than the vehicle can tolerate.

Depth is in some ways a limitation to a deep submersible. While *Trieste I* has claimed the greatest depth of the ocean in its historic dive there are still large penalties in choosing to operate at great depths. Most of these are weight/strength factors. It is impractical to design an all-purpose or all-depth vehicle since the heavy construction necessary for 20,000 feet or more makes it very costly to operate in shallow water as opposed to a shallow depth sub. So far more vehicles have been designed for the 2,000-

foot range than for any other depth. Over the last ten years the greatest activity in the ocean has been occurring along the world's continental shelves where the major fisheries, some minerals, and much of the oil reserves are to be found.

Most submersible operations have been plagued by the problem of reliability in the vehicle systems. The high failure rate of such things as underwater connectors, solid state inverters that change DC to AC, motors operating in oil, hydraulic leakage, severe corrosion, ground loops and short-circuits, and embolizing cables are in part due to the lack of adequate money devoted for development and also a general lack of knowledge of ways to design equipment for the undersea environment. As was proven in the space program, high reliability is expensive. While diving a submersible may not be comparable to a voyage to the moon, it only takes one tiny insignificant part failure to abort the dive or cause a costly delay. The lack of reliability in submersible systems has hindered a number of diving operations.

Over a period of ten years in which the modern manned submersible has been evolved we have seen this machine demonstrate its capability to carry out a number of jobs underwater more efficiently than had been done by surface ships. Next we will look at some of the specific accomplishments of the submersible.

# 7

# Submersible Vehicles– What They Have Done

By mid-1970, the beginning of the second decade of research submersibles, there were an estimated seventy vehicles in existence. Compared with cars, boats, or television sets, the number seems infinitesimal. Furthermore, with so few, it ought to be possible to know exactly how many there are, where each is, and what they are about in the way of diving. Yet each vehicle is a one-of-a-kind (excepting two or three small ones) and has been specially designed. Sometimes it was part of a company plan which was later scuttled in the interest of the management's profit and loss statement, and the vehicle has disappeared into the company warehouse or garage. Of the seventy submersibles, probably one-third are idle and in storage, perhaps never to swim again. Another group is finishing construction and sea trials, and preparing to go to work as designed. Finally, there is a relatively small number that have been most recently at work and have contributed significantly to scientific research or engineering and salvage efforts. It is this last group which deserves attention and evaluation. Details on individual specifications of all vehicles will be found in Appendix A.

Nearly all the activity in submersible vehicles occurred over a period of five years from 1965 to 1970. During this time, a relatively large number of dives were made. While several of the groups operating subs keep meticulously accurate diving logs and make these logs available to interested scientists, others, as a part of company policy, do not divulge the nature and number of dives made; still others make so many shallow repetitive dives that no records are kept. The practices of operation and record keeping are as varied as the vehicles. Thus far, no regulation exists governing this aspect of the submersible industry. As far as trying to sum up the number of dives made on all the civilian and military research submersibles is concerned, it is particularly difficult and subject to a wide range of estimates.

In 1970, Robert Ballard and K. O. Emery of Woods Hole Oceanographic Institution evaluated over 2,000 dives made for scientific research by thirty-five vehicles between 1950 to 1970. The majority of the dives were made with sixteen boats over a period of eight years. At least an equal number of dives have been made that were not for scientific purposes. Thus, it's reasonable to say that probably a minimum of 5,000 dives have been performed. These would add up to 10,000 hours of bottom time since two hours is an approximate over-all average dive length. Estimates by some scientists indicate it is possible that as many as 10,000 dives have been made in this five-year period.

At this point, it would be pertinent to consider what factors determine the value of a submersible. Some of the guidelines or criteria used by submersible engineers are: number of dives, length of dive, (effective bottom time, depth capability, number of observers carried, payload), amount of instruments and samples, speed and range, ability to carry out complex tasks, special-purpose tools, handling ability for launch/recovery, support requirements, and cost.

At this rather early stage in submersible development, it is difficult to compare these factors on various boats. The number of dives is certainly a significant factor. Generally speaking, vehicles that have made fewer than 100 dives have not accomplished much, while, on the other hand, boats that have made more than 500 dives are generally regarded as being successful. The factor of number of dives is tempered by the length of time spent on each dive. Vehicles that are able to spend less than two hours per dive cannot be very effective in carrying out complex tasks or bottom survey work. The range of bottom time of vehicles runs from several hours in the small, light boats to twenty-four to thirty hours in the larger ones. A few were designed to spend thirty to forty-five days submerged.

Another revealing test in evaluating the more successful subs is simply to look at those that have been on fairly continuous work assignments. The customers who pay for the use of the vehicles have gone through this evaluative process to decide where their research dollar or salvage dollar will be best spent. The largest customer or user group has been various branches of the U.S. Navy and some other government agencies. Vehicles that have been used the most by these were *Deepstar 4000*, *Alvin*, *Aluminaut*, and the *Perry* vehicles.

Finally, by reading the literature and the records of various vehicles, we can see the range and variety of tasks accomplished. There are relatively few vehicles that can perform the majority of the missions discussed in Chapter Six, "Work for the Manned Submersible."

Those that have best been able to do the work set out for them and have done it consistently will be discussed in some detail. Other vehicles that have had too little use to prove their value are briefly mentioned.

# Diving for Science

*Alvin,* the first American deep submersible, was built as a three-man vehicle. She is 22 feet long, weighing 29,000 pounds, and designed for diving to 6,000 feet for scientific missions. She was launched at Woods Hole, Massachusetts in June of 1964. There followed a period of several months of outfitting and modifying her to the requirements of the operating crew at Woods Hole. The designers had had little experience in submersible human factors and layout. They found that after a number of shallow dives in the harbor at Woods Hole, many changes were necessary in the location of instruments and controls. After the first season of local dives, *Alvin* was overhauled in readiness for the 1965 season and the necessary preparation for the final deep dive to 6,000 feet that would complete her sea trials. During the summer of 1965, *Alvin* was taken south aboard her ungainly and slow support craft *Lulu,* where she dived in the deep water around the Bahamas. After prolonged propulsion problems, *Alvin* made a successful manned 6,000-foot dive and became the first American submersible to obtain the Navy certification required to operate a Navy-funded vehicle. This meant she had passed the stiff examination given to U.S. Navy submarine hulls. *Alvin* was overhauled again during the winter of 1965, this time correcting many of the earlier problems. In February, 1966, she was called to active service as part of the search off Palomares, Spain for the hydrogen bomb lost at sea after the refueling accident of two United States Air Force aircraft.

*Alvin* was flown to Spain in two pieces. There she joined the other submersibles—*Cubmarine* and *Aluminaut*—in the search along with Task Force 65, the largest Naval effort ever assembled for an underwater object search. The area where the bomb was most likely to be found was deeper than 2,000 feet, a rugged terrain formed of foothills, steep slopes, and narrow valleys. Searching was slow. The submersibles were used to verify "targets" detected by a towed sonar. But the area was scattered with many false targets, so little seemed to come of this approach. Navigation among the steep hills and valleys was difficult. Sonar beacons and sonic devices were of little use because once the submersible was behind a hill the sound was obscured.

*Alvin* pilots Marv McCamis and Val Wilson finally discovered a likely looking trail leading down a slope at about 2,500 feet. The slope was so steep that *Alvin* could not follow it bow down since mercury trim could not lower the bow enough. The pilots estimated the slope at 70 degrees—an estimate exaggerated by the refraction of the port—but nevertheless very steep. Instead, *Alvin* was backed down the track, letting observers follow it through the side and bottom ports. At 2,532 feet, the bomb was spotted. The detection of this furrow made by the bomb as it slid down the ravine shows the value of visual observation from a manned vehicle. *Alvin,* on subsequent dives, managed to attach a line for a surface ship to lift the bomb. Unfortunately when the ship lifted the bomb, the line parted and the bomb, although harmless and not armed, plummeted down deeper into the labyrinth of ravines. *Alvin* went back to the painstaking search. She finally found the bomb again and patiently sat by on a "bomb watch" while *Aluminaut* also stood watch until a remotely operated cable-controlled device called *CURVE* managed to grab the bomb firmly and bring it to the surface for all the world to see. *Alvin* had performed admirably, making thirty-four dives totaling 228 hours in the water—several of them as long as eleven hours each. She had been the first submersible vehicle to attach a line for lifting by using her manipulator.

*Alvin's* operational crew from Woods Hole learned a great deal about navigating a submersible in rough terrain from these dives, which gave them the experience for subsequent dives along the East Coast slopes. The first of these later dives was off Bermuda for inspection of the underwater listening hydrophone arrays used by the U.S. Navy. While in the same area, a series of geologic bottom investigations were made for the Naval Oceanographic Office. During the following winter, *Alvin* was again refurbished. This time improvements were made in the viewing positions. The pilot used the forward port to see out and the scientists were given more space at each of the side ports. This arrangement made the best of a poor placement of view ports where the pilot and observer did not have overlapping views and thus were never sure of whether they were looking at the same thing. Cousteau had foreseen this difficulty and insisted on two forward ports with overlapping view angles—one each for pilot and observer.

Syntactic foam was added to the bow increasing the payload by 600 pounds. This allowed it to carry its new improved manipulator and tool rack at the same time. A superb suit of tools was devised by the engineers at WHOI.

During 1967, *Alvin*, operating from the catamaran *Lulu*, managed to make fifty-five dives—forty-two for the oceanographic scientists. The majority were for biology and geology. Dives made off Norfolk, Virginia showed biologist Bob Edwards, from the Bureau of Commercial Fisheries, that the submersible was an excellent means of observing ground fish communities. He was surprised at the good visibility even in shallow water and believed that studies of haddock spawning, territorialism of redfish, and diurnal or daily migration of hake all could be studied from such a vehicle. This was one of the dives that led to the use of other submersibles for fishery research.

*Alvin* also experienced some unusual events. While on a routine geologic dive on the Blake Plateau at 1,800 feet, an incensed, irate 200-pound swordfish took a strong dislike to *Alvin* and attacked her by striking his bill in the joint of the forebody-afterbody separation. *Alvin* survived, however, and the swordfish bill only penetrated the fiberglas skin. The swordfish was trapped and when *Alvin* surfaced the cook made it into steaks. This is the only reported fish attack, although *Deepstar 4000* pilots have seen broad-billed swordfish at 2,500 feet in the Gulf of Mexico. Bill Rainnie, Operations Chief and pilot of *Alvin*, expressed some concern at the possibility of such an attack, malicious or accidental, on one of the plexiglas ports. A port under high-pressure loading might not be able to withstand such a sharp force and could crack, catastrophically admitting water.

Another incident which proved that the WHOI group had learned a lot about search and recovery, as well as science, happened late in 1967. While diving 100 miles south of Martha's Vineyard, Massachusetts, *Alvin* surfaced into 8-to-12-foot seas and a blustery 35-knot wind. As she was being maneuvered in toward the opening of *Lulu's* catamaran hulls, one of the tending lines became fouled in the ship's propeller. *Alvin* and *Lulu* collided, breaking the latch designed to release the manipulator arm. The latch let go and the arm, brand-new that year, parted and sank in 4,500 feet of water. Although a good navigational fix was not available, the skipper managed to locate within a mile the spot where the arm dropped before leaving for port. After suitable preparation, *Lulu* and *Alvin* returned a month later, and with only three dives managed to locate the 6-foot long arm. A vital aid in location, as other submersible pilots have found, was a specially designed sonar search device. This sonar is called a CTFM for Continuous Transmission Frequency Modulation. It operates by sending out a signal and scanning the return displayed on a screen very similar to radar. An object several feet high can be detected several hundred feet away. In this case, there was some thought that the arm might be buried in the bottom, so a transponder, a sonic device that responds when triggered by a CTFM signal, was dropped in the suspected one-mile area. The search was conducted around the transponder in a search pattern. On the third dive *Alvin* pilots spotted the arm. For recovery they carried a 300-pound lead weight which could be dropped. When they picked up the arm with a simple fixed hook, they dropped the weight, thus maintaining equal ballast, and surfaced.

*Alvin* had managed to find its own arm and save the replacement cost of $50,000. It was a successful windup of a year in

145

which *Alvin* had made a variety of scientific and salvage dives that proved the value of manned vehicles.

During her life, *Alvin* has been used for nearly every type of marine scientific and nonscientific mission. She proved suitable for measuring the loss of sound in sediments on the bottom. With her, Woods Hole geologists explored many of the submarine canyons cutting through the slope along the East Coast. With *Alvin*'s sophisticated set of tools, the geologist could take gravity cores and make rotary cores in hard rock. The biologist rigged a series of nets that could be opened and closed and used them to sample planktonic and slow-moving organisms.

*Alvin* had made some 307 dives by October 16, 1968 when it seemed it might be her last day of a short and happy life. During launching, with two scientists inside her and the pilot in the conning tower with the hatch open as was the routine procedure, a cable supporting the platform on which *Alvin* was being lowered broke * and she plunged on her side into the sea 18 feet below and disappeared beneath the surface. The two men inside were able to scramble out the hatch as she bobbed up to the surface for a brief moment before great torrents of water flooded in the open hatch, sinking her in 5,000 feet of water. Fortunately, no one was injured in this first serious submersible accident.

Efforts to salvage her that fall were useless, as stormy winds and high seas beset the New England coast. *Alvin* rested on the bottom during the winter and the following spring. The next chapter in *Alvin*'s career belongs to *Aluminaut*.

*Aluminaut*, the five-man, 51-foot, 83-ton all-aluminum submersible started her diving career in 1964 in Long Island Sound off Connecticut where many of the U.S. Navy's fleet of submarines started. After preliminary sea trials off Groton, she was towed to her new home base in Miami, Florida. In August of 1965, *Alumi-*

---

* Remember Cousteau had said, "You can trust a cable to do one of two things at sea—either get fouled or break."

146

*naut* made a series of dives off the Grand Bahama Banks to a depth of 2,750 feet for a total submerged dive time of thirty-three hours. This was a far longer duration than that of any of the existing vehicles and showed that *Aluminaut* was in the class of the larger vehicles, apart from *Alvin, Soucoupe,* and *Deepstar*. She wound up her first trials by sailing from the Bahamas to Miami across the straits of Florida—submerged at 1,250 feet on a 70-mile, ten-hour trek. Her speed was reported up to 3½ knots at times.

Next she dived to 6,250 feet for her deep test dive. Although *Aluminaut* was designed for 15,000 feet, she could not dive that deep; the hemi-heads—hemispherical ends of the aluminum hull—had shown some flaws after manufacture. These were machined out removing a half-inch of metal and thus compromising the ultimate strength. Reportedly, *Aluminaut* can dive to at least 8,000 feet with the present hull, at which depth there is a safety factor of 2.8. There was talk of finishing new heads to allow for a dive of the 15,000-foot depth, but the cost was never justified.

The owners and operators of *Aluminaut*—Reynolds International and Reynolds Submarine Services—decided that the scientific world had to be shown how many things could be done by a vehicle of its size and speed. Art Markel, Vice-President and General Manager of Reynolds Submarine Services, organized a series of dives off Miami for scientists from the University of Miami, U.S. Coast and Geodetic Survey, and the U.S. Navy's Deep Submergence Systems Program. In this series, *Aluminaut* showed that she had a sizable payload capability, could accommodate three scientists at a time, give them ample room to move around in, let them observe, and use the two large manipulators. The scientists were fascinated with being able to see for the first time what they had previously only suspected. This same revelation was experienced by West Coast colleagues using the Cousteau *Diving Saucer*. Their appetites were whetted. The demonstration

# Diving for Science

dives had accomplished what was intended. Dr. Conrad Neumann, a marine geologist with the University of Miami, wrote later of his first dive in *Aluminaut:*

> Nevertheless . . . with no sampling devices other than cameras aboard, I was very impressed with the great potential that this type of vehicle holds for future research. The entire bottom environment can be observed at once, not just glimpsed in patches as with a lowered camera. Processes can be observed *in operation.* . . . As a result of this experience, my opinion of deep diving vehicles has reversed from one of negative skepticism to enthusiastic optimism. I feel the manned submersibles are the best thing to come along in oceanography since galvanized wire and electrician's tape.

When this testimonial was published in *Geo-Marine Technology,* a technical journal, it helped to spread the word among ocean scientists that the submersible was worth using as soon as a set of tools and sensors could be fitted to it. As Dr. Fred Spiess of Scripps said of the 1964 *Diving Saucer* dives off La Jolla, "One dive equals one convert." *Aluminaut* was a good boat for making converts. It was roomy, had long endurance, reasonable speed, excellent manipulators, and provided good viewing.

But before they could get to that sort of work, *Aluminaut* was drafted, as *Alvin* had been, in February of 1966 for the H-bomb search. She performed well, especially during the "bomb sitting" detail after the lift line had broken and the Navy wanted to be sure the bomb did not slide farther down the slope. *Aluminaut* spent a 22-hour vigil, while *Alvin* had to surface and charge batteries and replenish life-support materials. *Aluminaut* was not suited to the rugged countryside of the steep slopes off Palomares and was not able to maneuver as well as *Alvin.* One noteworthy feature was that *Aluminaut* was large enough to carry a side-looking sonar for searching. This instrument had large, heavy transducers that did the sonic looking to either side as the sub moved along. A paper recorder inside showed a plan view presentation of all objects, rocks, and ridges that were reflecting the

148

sound. This sort of search tool aboard a submersible was valuable because investigation of likely targets was easier than by a surface-towed sonar and a separate submersible. A primary clue in the search was found with the side-looking sonar when crew members detected the tail section of the B-52 that had been carrying the bomb. This led the entire search into deep water and put *Alvin* on the trail of the bomb.

Over the years that followed, *Aluminaut* made some scientific dives but probably made more for salvage and engineering evaluation since she had more power and could carry a greater payload than any of the rest of the available submersibles.

On a dive off Miami, made in 1967, Dr. Conrad Neumann and Dr. Mahlon Ball of the University of Miami made several worthwhile observations. *Aluminaut* had been outfitted with a gravimeter—a sensitive instrument that can measure the gravity at a particular point. While the stability of the vehicle is important for gravity measurements, in this instance the readings were taken while the boat sat firmly on the bottom for twenty-minute periods. These readings were more accurate than previous ones made near the surface. By getting close to the geologic structure he was interested in, Dr. Ball was able to separate the effects of the overlying water from the observed gravity anomaly. Thus he found supporting evidence that the Florida Straits are a result of structurally downfolded or downfaulted light rocks—a fact not possible to determine from the surface.

"If we made no other measurements," wrote Neumann of the expedition in *Oceanology International*, "The value of these gravity readings obtained with no new or specially designed equipment would have 'paid the rent' and made the trip worthwhile."

On this ten-hour dive, they made many other observations. Among these, Neumann watched the current direction and behavior for over eight hours near the Miami side of the Straits, and noted to his surprise that the current flow was to the south —the opposite direction of the Gulf Stream. Although the existance of this counter-current sneaking under the strong northerly flow of the Gulf Stream had been suspected, it had never been proved. Most oceanographers believed a southerly component was due to tide effects. Using some dye and a stopwatch, the scientists aboard *Aluminaut* were able to determine the current speed and direction visually and spent long enough time observing to discount the theory that the tides would change direction after six hours.

*Aluminaut* was used by engineers from the Navy's Deep Submergence Program to test and evaluate relatively large and heavy sonars that would sink an *Alvin-* or *Saucer*-sized vehicle. The original and improved versions of the CTFM forward search sonar were evaluated on *Aluminaut,* as was a doppler sonar used to determine the speed of a submersible along the bottom. The test range set up at the 2,200-foot depth showed that more developmental work was needed on the sonar before it would perform as forecast by the manufacturer. This sort of diving only kept *Aluminaut* partially occupied, but it proved the need for a vehicle to do this sort of test work.

One of the few links from the rich heritage of past submarine diving in America was made when Art Markel and Reynolds designed a set of wheels for *Aluminaut* to be used for travel along the bottom. Harking back to Simon Lake and his *Argonaut,* *Aluminaut* was equipped with an undercarriage on which three rubber wheels could be bolted. The wheels let the vehicle travel at a uniform height above the bottom where the bottom was firm enough. Considerable publicity was gained for *Aluminaut* when she rolled down a "pavement" of black manganese-rich rock along the Blake Plateau off Georgia. She returned with a sample of the manganese-iron material weighing 168 pounds—one of the largest ever recovered. This area was mapped by the U.S. Geological Survey as having large deposits of low grade manganese-iron and phosphate-bearing materials that could be economically mined in the future.

# Submersible Vehicles—What They Have Done

Some of the other dive accomplishments of *Aluminaut* have included fishery surveys, the recovery of an expensive oceanographic current meter array, and the much heralded salvage and recovery of *Alvin*.

Diving in *Aluminaut*, over two days off Cape Kennedy, Florida, the Bureau of Commercial Fisheries' scientists discovered the presence of a large, but until then barely suspected, scallop fishery. In two ten-hour dives, scientists measured and photographed the density and physical layout of this bed—a task that would be quite difficult by trawling from the surface. As a result, a thriving sea scallop industry was started in this area several years later.

While operating for the United States Naval Oceanographic Office near St. Croix in the Virgin Islands, the *Aluminaut* crew was sent to locate and retrieve a lost array of six current meters. After some searching for about an hour, they found the white nylon line leading to the current meters lying on the bottom at 3,100 feet. On the next dive, *Aluminaut* carried a spool with 3,300 feet of nylon line. This end of the nylon was attached to the line of meters using the two manipulators, and the vehicle ascended, paying out line until the surface support ship was able to take the line. The six current meters, weighing 3,150 pounds, were retrieved and found in salvagable condition. In 1966, this was the deepest and possibly the only retrieval of lost instruments made primarily by a manned submersible. The ability to use both of the manipulators as a person uses two hands to perform a complex job was important to the success of the recovery.

Finally, *Aluminaut* played a valuable part in the salvage of her sister submersible *Alvin* in 1969. *Alvin* had been resting on the bottom since the previous fall, while the salvage group at Woods Hole waited for the calmer summer weather and an answer to how to bring the submersible up from 5,000 feet. During the spring, the highly instrumented survey ship USNS *Mizar* had located *Alvin* after searching with a towed camera and sonar fish.

151

Photographs showed *Alvin* sitting upright, hatch open, and relatively undamaged. The tail section with the steerable propeller had been broken, but not separated, probably when it hit *Lulu* as it was thrown into the water. Amazingly there was no silt or any evidence of particle fallout on the flat surfaces of *Alvin*, indicating an absence of sedimentation at that depth.

The *Alvin* operation was supported by the Navy, who had originally planned to do the salvage operation itself. The Navy Salvage Office had gained a great deal of experience with the H-bomb and the *Thresher* search. The men in charge contemplated using the latest version of the cable-operated device, *CURVE III*, similar to one used at Palomares, Spain.

But for once free enterprise prevailed. Art Markel proposed to take *Aluminaut* to Boston and salvage *Alvin* on a "no cure, no pay" basis. If *Aluminaut* couldn't salvage *Alvin*, then Reynolds would not charge the Navy for the effort. The Navy accepted, and in July, 1969, *Aluminaut* was loaded aboard an offshore service ship, the *Stacey Tide*, for the trip to Boston. From Boston *Aluminaut* was towed to the diving site and the salvage commenced.

*Alvin* had two lift points where she could be picked up by using a spreader bar. There was no easy way to make such a connection, so it was decided to insert a toggle bar in the open hatch. In this way the lifting strain would be spread over the hull. The plan was simple. *Aluminaut* would carry a large spool of 4½-inch nylon line with the toggle bar at the end. She would insert the toggle bar in the hatch and unreel the line as she surfaced. Unfortunately this plan did not work because rough seas prevented the attachment of the cumbersome line reel to *Aluminaut*. As with all good salvage plans there was an alternate plan available. USNS *Mizar* lowered the toggle bar on the salvage line to the bottom very close to *Alvin*. *Aluminaut* maneuvered into position with the bar held in her manipulator arms. In this position, *Aluminaut* looked like a giant praying mantis. The toggle bar went through the open hatch and wedged neatly inside across the opening. Then *Mizar*'s winches hauled *Alvin* to within 85 feet of the surface where a net was placed around her to make sure nothing dropped off in lifting. *Alvin* was then towed to shallow water. It was September, nearly a year after sinking, when she was brought on board the salvage barge. The *Aluminaut* crew was triumphant since they had "cured." *Alvin* was quickly hustled to safe quarters where corrosion inhibitors were used to prevent rapid deterioration from exposure to the air.

Although *Aluminaut* has not made as many dives as some of the smaller vehicles which may make two or three dives in a day, the dive duration of the larger boat has been much longer. The average dive length of *Aluminaut* is nearly ten hours, giving it a total of over 2,000 hours underwater.

*Deepstar 4000*, the three-man vehicle for 4,000-foot depths, completed by Cousteau's group in Marseilles, was given several quick shallow dives in the harbor and then flown to Annapolis, Maryland. Here, a minor overhaul was done to Americanize fittings and modify the vehicle for use by Westinghouse's Underseas Division. A series of sea trials was staged in Chesapeake Bay, but the brackish warm water had a too-low density and made buoyancy control difficult. *Deepstar* was taken to San Diego, where, over a long arduous period of eight months, she went through sea trials and Navy certification. Westinghouse learned, as other operators before and after, that sometimes the debugging may take an unreasonably long time, far beyond that planned on. In this case, the support ship and crew were standing by at a cost of $90,000 per month.

On June 4, 1966, *Deepstar* became the first submersible owned by an American industry to be certified by the United States Navy. On that date, she dived to 4,132 feet off San Diego.

*Deepstar* had an easier beginning than *Aluminaut* since the momentum created by *Diving Saucer* in 1965 had resulted in a group of scientists with well-prepared plans eager to dive. The program was ambitious since the Navy scientists contracted for 193 diving days in the nearly thirteen-month period from June 6 to July 2, 1967. A diving day was termed any day the vehicle was ready for diving. It also included time necessary for the ship to reach the diving site, as well as days lost to bad weather. Westinghouse agreed to provide *Deepstar,* an operating crew, a ship, and all maintenance for the 193 days for a price of $1,200,-000. The Westinghouse group had learned from experience with the *Diving Saucer* that agreeing to a quota of dives was difficult to meet and provide. As it was, meeting the 193 operating days was not an easy task—even for a well-trained team.

NEL had the majority of the dives lined up. A sister Navy laboratory, Underwater Sound Laboratory (USNUSL) of New London, Connecticut had agreed to pay for about one-quarter of the nearly 200 dives. USNUSL dives were spread out from the Gulf of Mexico to Nova Scotia. This meant that *Deepstar* and crew traveled from San Diego to the Gulf Coast and then along the entire Atlantic coast and back to San Diego—diving along the way. Since most of the NEL scientists had dived before, in *Trieste I, Trieste II, Soucoupe,* and the French bathyscaphe *Archimede,* they knew how to go about diving for science. Their various experiments all involved elaborate oceanographic equipment fabricated at the NEL shops. Some of the gear was quite heavy and required special mounting that would permit jettisoning in emergency. *Deepstar* had about a 500-pound payload using the three-man crew of a pilot and two observers. If the instruments such as corers, water bottles, probes, velocimeters, and current meters all had to be mounted on the bow, there would be far too much weight carried too far forward. A rather clever way around this was devised by NEL and Westinghouse engineers. An instrument brow was fabricated. This was simply an aluminum framework that projected out five feet and was attached to the bow. To offset the added weight forward, a number of blocks of syntactic foam were stacked on the

brow to give increased buoyancy. Another 450 pounds of payload were gained this way as well as redistribution of the load.
To make the brow safe, explosive bolts were used in attaching it
to *Deepstar*. If the whole works had to be dumped in an emergency for extra buoyancy, the pilot had only to throw a switch
and the seven bolts would shear explosively, releasing the instrument brow.

On the first set of scientific dives off San Diego, John Beagles
tried to determine more about sediment transport over the slope
west of the NEL Oceanographic Research Tower. He moved in
*Deepstar* from 50 to 500 fathoms on a traverse that showed a
rather sparce benthic animal life except for solitary corals, sea
anemones, and sea pens. The movie camera mounted inside between the pilot and observer, just as on the *Soucoupe,* was used
to obtain motion pictures of the down slope journey. Still photographs also were taken. *Deepstar* had one of the best photographic arrangements of cameras and lights of any of the existing vehicles because Westinghouse realized the importance of
good photographs for the submersible scientist.

On these earlier dives the brow had not yet been completed
and John Beagles took along a small La Coste-Romberg gravimeter for use when *Deepstar* sat motionless on the bottom. Ten
stations were occupied in water to 3,000 feet in depth—perhaps
some of the deepest gravity readings ever taken. He discovered
an important point, however, when he tried to interpret the
gravity readings. Although he could read the results from the instrument to 0.013 mgal.—well within accuracy requirements, the
exact position of *Deepstar* was not known well at all. To be of
any value for gravity determination the vertical position (i.e.
depth) had to be known to the nearest 3 feet and the horizontal
error (i.e., longitude and latitude) could not exceed 62 feet. In
all likelihood, *Deepstar's* position was probably ten times in error
of this figure since her depth was only readable to within 15 to
30 feet, and the relative position of *Deepstar* and her support

ship were loosely controlled probably at best to several hundred yards. The problem of precise positioning was one that needed considerable attention before this sort of precise survey work could be attempted. In 1967, the Woods Hole group was already well aware of the problem and taking positive steps to improve *Alvin*'s capability. The *Deepstar* team was less interested since their contractual commitment called for availability of the vehicle—not for positional accuracy. While *Deepstar* was superior in many ways to previous submersibles, the need for positional accuracy was neglected for some time.

By the time the instrument brow was certified by the United States Navy as safe, Dr. Gene LaFond had developed a rack of ten Fjarlie bottles for sampling water at any desired spot. These metal tubular bottles were mounted horizontally out in front and the opening valves were controlled by Dr. LaFond from inside. A Savonius rotor current meter was rigged so it could be "planted" and the vehicle could back away while still connected by an umbilical cable 50 feet or more long. A water temperature sensor and a bottom sediment temperature probe were attached to the brow. Finally, the cameras were aimed to photograph the equipment and any animal life seen. The purpose of this dive and. many more like it to follow using this equipment setup was to measure and observe the physical and chemical properties of seawater that affected sound velocity very near the bottom. The microchemistry, thermal effects, and water motion on the sea floor are practically impossible to study from a surface ship using conventional "over the side" techniques.

The first series of dives were plagued with equipment failures as might be suspected with a lot of newly integrated gear. Nevertheless, Dr. LaFond found some interesting variations in the salinity, phosphate, nitrate, and silicate contents within 40 feet of the bottom. In San Pedro Basin, off Los Angeles, temperature probes at 3,000 feet showed significant increases in bottom sediment while the current flow was pulsating with speeds to 0.2

knots. Eleven dives of this type were made over the next few months. They were among the first instrumental dives of a submersible for physical oceanography in microscale like this.

Another type of dive in which *Deepstar* carried on the tradition of *Soucoupe* was for biologists interested in detailed study of the Deep Scattering Layer (DSL). One of the world's leading authorities on the identity and behavior of the tiny marine organisms that compose the DSL is Dr. Eric Barham of NEL. Dr. Barham made earlier observations of horizontal and vertical movements of plankton and fish and their daily migrations from depths of 350 fathoms to the surface using the *Diving Saucer* in 1965. The technique established by biologist Barham was first to survey the area with an echo sounder of the surface ship which showed the position of the organisms since they act as sound scatterers. The dive consisted of a series of transects in which *Deepstar*'s lights were turned on every three minutes for one to two minutes for a "light look." Interspersed were two-minute dark periods. This enabled the scientists both to see and to photograph the animals and also to move slowly down during the dark periods and not "drag" the animals attracted to the lights. After making several dives both during the day and at night, Barham found that in general the upper layer of the two-layer system in the San Diego Trough, 30 miles at sea, showed a majority of hatchet fish while the lower layer was composed of siphonophore jellyfish. Both animals appeared to be able to scatter sound but the fish were believed by Barham to be the primary scatterers.

*Deepstar* was easy to maneuver because it had good depth control that let it remain nearly motionless in the water column and thus the scientists were able to look at the animals from only a few feet away without disturbing them. During the third such dive, Barham and *Deepstar* hung at 400 feet for over an hour turning lights on and off and photographing.

During this first operational period, *Deepstar* made 101 dives for fifty-three scientists between June and December, 1966. Some of the other dives were for determining the shear strength of bottom sediments, measuring *in situ* sound velocity of sea floor sediment, measuring the gamma radiation from surface to 4,000 feet, and measuring the sound velocity of the water. The results were of value to the scientists although not always of great excitement to the layman.

Several dives were more than routine. On one in particular, Gene LaFond was the diving scientist and the pilot was Joe Thompson, one of the original *Deepstar* crew who had trained on *Diving Saucer*. They were on one of Dr. LaFond's regular dives to measure physical and chemical properties on the bottom and the heavily rigged brow was on *Deepstar*. Joe Thompson had just maneuvered backwards after carefully placing the Savonius current meter on the bottom with the mechanical arm. His attention was focused on moving gently away a short distance and waiting for any turbulence to settle down before recording the faint pulsating currents. The bright lights illuminated the bottom at 4,000 feet in the San Diego Trough, surrounded by a black perimeter. Suddenly an enormous fish swam into the space under the port and Joe saw the eye, which he described as the size of a dinner plate. The fish moved by with a swirl of fine silt off the bottom. Finally as Joe watched speechless the serrated tail of the fish passed by. LaFond from his position could only see the swirl of mud. Joe judged the fish to be at least 30 feet long—maybe 40 feet. Since they were attached by cable to the current meter, Joe could not turn the vehicle to follow this monster fish. Nor was there time or presence of mind to get a photograph—it all happened too quickly. Later they discussed their fish story with experts in marine biology. Dr. Carl Hubbs from Scripps, for one, thought this might have been a Greenland shark. Dr. Hubbs had taken photographs of these 20-foot beasts by lowering cameras in great depths where the water is cold. But Joe Thompson still believed it was a true fish because he was sure it had scales—and a shark does not. The Thompson-LaFond "sea monster" has neither been identified nor ever seen again. It confirms what some of the more imaginative persons associated with marine life have always wanted to prove—that there are real sea monsters.

*Deepstar 4000,* a little like her colleague *Alvin,* has had several incidents that proved its sound design by the French builders. None of these were as serious as the *Alvin* loss, but one, Dive 162, was termed by the Navy a triple casualty. The pilot again was Joe Thompson, veteran of many *Deepstar* dives. The passengers were Dave Moore and Gene Lewis. This dive was made close to San Diego in 3,750 feet of water. The descent was normal, or it appeared to be. What Joe didn't know was that the water ballast system composed of some eight small bottles had flooded. These bottles were normally empty at the surface. At any time, the pilot could activate a valve admitting a shot of water, making the vehicle heavier. Instead of being able to pump the water out as with most ballast systems, there were fifty-two small one-pound weights that could be released individually, thus offsetting the water weight and lightening the boat. Approaching the bottom, Joe had let go the 220-pound descent weight to slow the descent of *Deepstar.* As they landed, he felt that the vehicle was heavy. He turned on the propulsion to move ahead, but found that they were dragging on the silty bottom. At first, he thought that perhaps the descent weight located on the after end of the boat had jammed and was still on. There was no way of telling. Next, for no related reason, the inverter that supplies AC to the hydraulic system failed. This was the second casualty and meant that Joe could not pump mercury to change the attitude of the boat, or drop the 150 pounds of small weights since they were released hydraulically. He had landed with the bow 30° up to protect the scientific gear. Due to this high bow angle, when he released the 187-pound weight to return to the surface, the rectangular lead weight caught one corner on the weight box and would not drop. The vehicle was unable to surface. The men sat quietly at 3,750 feet beneath the ocean to decide calmly what was best to do next. *Deepstar*, like most of the rest of the major vehicles, has several emergency of back-up systems.

After relaying the details by underwater telephone to the surface ship, *Burch Tide*, Joe proceeded to jettison about 200 pounds of mercury into the ocean, costing about $13 per pound. Nothing happened. They were still too heavy to overcome the weight of the water in the flooded ballast bottles. He tried using the propulsion motors to drive the boat up but only succeeded in skidding along the bottom, picking up mud and making the boat heavier. There remained two things to do—drop the forward battery case weighing 600 pounds in water, or drop the scientific equipment on the brow. Joe chose the battery. He turned the handle directly above the ports that pulled the cable which would drop the case. This maneuver had been tested with a simulated battery at certification time. With a great whoosh and shudder, the battery case—5 feet long—came crashing out and down onto the bottom. The jolt also released the jammed ascent weight. *Deepstar* rapidly took off for the surface and arrived only a few minutes later—minus battery and mercury but with its precious cargo of three men.

The following three months were devoted to a major overhaul of *Deepstar* in preparation for her extensive tour of the Gulf and Atlantic. Static invertors were substituted for the heavier rotary ones built in France. A new, variable ballast oil system that could not flood was installed. All the welds in the hull were x-rayed for metal fracture or fatigue. In April the crew was transferred to New Orleans. *Deepstar* was moved aboard a brand new mother ship called *Search Tide*, a 165-foot ship with 130 feet of deck space. The 20-ton crane and living and support vans from *Burch Tide* were all placed aboard the new ship.

This whole mobile diving rig then went from Florida to New England for the Underwater Sound Lab making approximately sixty-five dives through November of 1967, and returning to San Diego by way of Mexico and Panama in the Spring of 1968. By 1969, *Deepstar*, during the contract to the United States Navy, had made a total of 504 dives in three years. Results of these dives would be appearing in papers by scientists for several years to come.

Several other vehicles have carried on diving activities during this period of 1966 to 1970. First, let's consider some of the larger submersibles.

The 40-foot, 52-ton vehicle, *Deep Quest,* designed and built by Lockheed Missiles and Space Division, holds the depth record of 8,350 feet for an American-built submersible. With this entry into deep submergence, Lockheed became qualified to build the U.S. Navy's *DSRV. Deep Quest* is a complex boat, well instrumented and more sophisticated in control and navigation than any of the other commercial vehicles. She has a specially designed surface support ship, *Transquest,* a 105-foot vessel able to handle *Deep Quest's* heavy weight and large size. The sub is best qualified for extended search and salvage and in 1969 she did an excellent job of finding the flight recorders of two commercial airliners that crashed at sea off Los Angeles. Like *Aluminaut,* she has a large lift capacity and can pick up at least 2,000 pounds. Lockheed outfitted *Deep Quest* with television, video tape recorder, cameras, and a multiple-coring device. *Deep Quest* has not seen a lot of diving service; of approximately 130 dives, only forty have been for scientific research.

Professor Auguste Piccard had the idea for the mesoscaphe or mid-water boat, and passed the preliminary design of this to his son, Jacques. The mesoscaphe was meant for long period missions in between the surface and bottom. After the death of his father, the younger Piccard obtained the backing of Grumman Aircraft Corporation to produce a mesoscaphe capable of drifting with the Gulf Stream. *PX-15* was built in Switzerland in 1967, and was brought to this country in 1968 to be christened *Ben Franklin.*

161

She is a 48-foot, 130-ton, five-man submersible designed for a maximum depth of 2,000 feet. Her principal feature is that she can remain submerged for forty-five days due to a specially designed life support system modeled after those used in space craft. During the summer of 1969, *Ben Franklin*, under command of Jacques Piccard with four oceanauts, drifted 1,440 miles submerged in the Gulf Stream. During the thirty-one days these scientists were able to learn more about the total behavior of the stream and many eddies. *Ben Franklin* has been used for a few benthic observation dives since then and has completed about seventy dives, although her diving hours total over 1,200. Like *Aluminaut*, her size makes her difficult to transport over long distances as opposed to the smaller *Deepstar* and *Alvin* types which are small enough to bring aboard a ship.

*Deep Ocean Work Boat, DOWB*, was built by General Motors as a scientific obervation vehicle. In many ways it is a second generation of *Deep Jeep*, a U.S. Navy experimental submersible built in 1964. *DOWB* is a two-man vehicle, 17 feet long, and weighs 9 tons. She has strong emphasis on top and bottom 360° visibility using a set of optics rather than operator view ports. The operators sit upright and look through eyepieces. *DOWB* has made some salvage and recovery dives to 6,300 feet in a weapons range off Santa Catalina Island, California. In 1969 she attempted to find *Alvin*, but malfunctions and winter storms prevented any chance of success. She became fully operational at a time when money for diving was short. *DOWB* has made about 120 dives.

International Hydrodynamics of Vancouver, British Columbia has built four submersibles along similar lines. The first, *Pisces I*, although originally designed for 5,000-foot depths, has been re-rated to 1,800 feet after several years of logging over 625 working dives. This vehicle was built on a minimum budget, using conventional techniques such as shifting batteries fore and aft instead of using mercury for attitude control. *Pisces I* uses an air ballast system rather than droppable weights and thus can make a number of dives without coming out of the water for weight reloading. She has been dived in the Arctic ice areas, used to recover spent torpedoes in Washington's Dabob Bay, and made some fishery trawling observations. Using one of several manipulators that operated a hydraulic chain cutter, *Pisces I* managed to secure a lift line to a 90-ton tug in 670 feet of water so that it could be salvaged. This was a record for salvage of a ship this size. *Pisces III,* similar in design but with a depth range of 3,500 feet, has made over 300 dives doing work in the Pacific Northwest and Hudson Bay.

John Perry, one-time newspaper magnate in Florida, built the original Perry *Cubmarine (PC-3X)* in 1960, and has since expanded his small company into a factory that has turned out at least eight vehicles of differing designs. In terms of numbers, Perry has probably been more successful than any of the other companies engaged in submersible construction. The early boats were two-man, 2-ton, 300- and 600-foot depth designs that have collectively made over a thousand dives. Two of the 300-foot *PC-3A* subs have been used in the shallow lagoons of the Marshall Islands by the United States Army and United States Air Force for missile component recovery. These vehicles, with many ports around the side, are best suited for observation in shallow water. Two deeper boats were built for continental shelf depths. *Deep Diver* was the first of the modern commercial vehicles to have a swim-out lock, and she made a dive to 700 feet when a man spent fifteen minutes outside collecting specimens. *Shelf Diver* was similar, but instead of 1,300-foot maximum depth, she can only reach 800 feet. While the Perry boats lack the sophistication of the big company multimillion dollar vehicles, they have been versatile in shallow water and quite instrumental in getting scientists and engineers started in using the submersible.

It should be evident that the manned submersible has made many contributions to the scientific record of oceanographic research. The results from the 2,000 documented scientific dives have been published in some 350 papers and articles between 1950 and 1970. Nearly half of these articles came from dives made by American scientists. Of the total number of dives of vehicles of all countries, the majority were for geological missions, with biological dives a close second.

# 8

Submersible Safety,
Testing, and Certification

A certain amount of safety has been built into
most submersibles ever since the first ones were operated. One
of the prime functions of an undersea boat was to take persons
beneath the sea for one of a number of reasons and safely return
to the surface. Only recently has there been a systematic and
relatively thorough approach to assure that all submersibles will
be safe.

There are several outstanding examples of submersibles and
submarines which were extremely unsafe and resulted in a loss of
life as well as the vehicle. The earliest was probably the unfortunate "sloop" design of John Day which lacked a ballast release
system, thus being unable to return to the surface. Later on, the
almost comic tale of the *Hunley* and its numerous sinkings because of instability, lack of life support equipment, inadequate

crew training, and lack of over-all mission plan of action is the classic example of disregard for the basic elements of submarine safety. Three crews of ten men were killed because the vital elements of safe design and operation were not followed.

The most fundamental elements usually included in determining submersible safety are the material condition of the submarine, crew training and operational maintenance, and an operational plan for vehicle use.

The Confederates' *Hunley* violated all three of these safety elements. The design was not safe because the vehicle lacked longitudinal stability and tended to dive wildly. There was inadequate operator training. Finally, there was no over-all operation planning since the use of a spar torpedo close to the *Hunley* destroyed both the *Housatonic* and the *Hunley*.

Of course, it can be argued that this was over one hundred years ago and we now have a far better approach to safety on underwater craft. While it is true there have been very few disasters with American submarines, the loss of the *Thresher* and *Scorpion* have caused the U.S. Navy to institute an improved safety program. This program, called SUB SAFE, includes a rigid and specific analysis of design, material, and construction.

The safety of the small manned submersible to be used for scientific missions has been less easy to define and initiate. Certainly for the first half of the 1960–70 period, there was little uniformity in the elements of design, materials, and construction. Few rules or regulations, or even safety guidelines existed for the vehicles already built or for the twenty or so that were under construction in 1965. Although there was no definite set of rules, most of the companies and groups involved in building a submersible sought to insure having a safe vehicle by starting with the design. In general, the parts of a submersible that can directly affect safety are: pressure hull; flotation and buoyancy systems; emergency weight release devices; life support systems; emergency power, breathing, and sonic locating devices. Each of these systems or components has to be designed to function in the environmental conditions normally found. Usually a margin is provided for abnormal conditions such as high attitude, power failure, extreme cold, and so on. Most companies make it a practice, during the stages of preliminary and final design, to hold formal design reviews so that more than one or two design engineers get a chance to make things work on paper. Many times what appears to be an ingenious and novel approach in a design can be "shot down" by pilots and operators who have the experience of operating the vehicles at sea.

After a safe design has been obtained, the vehicle and its equipment must be built and assembled with precision and care. This involves the quality of the materials and it is the responsibility of the quality control engineer to know that each component meets the required specification. One of the most positive ways has been to test each component under operating conditions. This requires elaborate testing devices, pressure chambers, considerable time, and manpower. But if the tests are done correctly, chances are good that the submersible will perform reliably and safely.

Knowing that a thorough job of design and fabrication has been done is one thing; proving it to the certification agencies is another and requires thorough documentation in the form of drawings, calculations, and records. These procedures are done quite differently to meet the requirements of the four different agencies involved with the certification of noncombatant submarines or undersea vehicles. These agencies are the U.S. Navy, American Bureau of Shipping (ABS), U.S. Coast Guard, and the Marine Technology Society (MTS).

Some confusion exists over the term certification as applied to submersibles—why some are certified by the Navy, others by ABS, and a few are operated without any type of certification. Why is this and does it mean that some are safer than others? Or, if a vehicle is not certified, is it unsafe?

167

The agency that has been certifying submarines safe since the inception of the submarine is naturally the U.S. Navy. A detailed and precise set of rules and procedures exists for military submarines. The certification for noncombatant deep diving vehicles was altered to delete those tests unapplicable to research diving vehicles and to include a lot of new items. In 1966, *Alvin* became the first American DRV to be certified by the Navy. Certification meant that *Alvin* was safe for naval or government-paid personnel to use for the specific research missions it was designed for. At the time no "book" had been written and the process was slow, taking nearly a year for *Alvin* to become Navy certified. It was especially difficult because *Alvin* was already built which made it harder to document the material used and the steps taken during fabrication.

*Deepstar 4000*, built in France, was also Navy certified in 1966 and was the first industry-owned vehicle to pass the Navy's requirements. Subsequently, a well-spelled-out procedure was issued by the Navy which defined certification. Two volumes were published in 1968 (see Bibliography). A few other vehicles besides the first two have been certified. The process is quite thorough, time consuming, and relatively expensive. It is usually only performed on those vehicles the Navy owns or intends to contract for use. Each certified vehicle must be recertified each year or when beginning a new contract for a Navy-funded group. The average cost for Navy certification is $25,000. So, if a Navy laboratory wanted to use a privately owned vehicle for a week or two, it would hardly be worth the certification cost and time.

But not all vehicles need to be Navy certified. Those operating in the civilian field can obtain certification from the American Bureau of Shipping. ABS was requested to set up a special submersible committee by the Navy. As a result, the book, *Guide for the Classification of Manned Submersibles,* was published by ABS in 1968. Like the Navy's two books, it set forth specific design criteria, examination procedures, and tests for research submersibles. ABS is primarily interested in the fabrication of the hull, while Navy certification covers not only the hull materials and fabrication, but crew training and mission plan.

A third group that has been actively instrumental in establishing a set of safety guidelines is the Marine Technology Society. MTS is a professional society formed by engineers, scientists, and government employees to further marine affairs. While it in no way acts as a certifying body, MTS has been a leader in issuing what is probably the most comprehensive publication on the submersible and its safety; published in 1968, it is called *Safety and Operational Guidelines for Undersea Vehicles*. If the suggested guidelines are followed, it is likely the vehicle will be a lot safer than if they were not followed. The MTS book does not have any inspection criteria as the Navy and ABS regulations have.

The fourth agency interested in the regulation of submersible activity is the United States Coast Guard which has had responsibility for the safety of surface craft and occupants. In 1971 the Coast Guard was still awaiting the passage of legislation by Congress which would define its mission regarding the submersible. Once this legislation is enacted, the Coast Guard will establish rules that must be followed in order to license or operate a submersible working out of a United States port. There will be several classes of vehicles such as passenger carrying, cargo, repair and industrial work boat, and experimental or research craft. Until this regulation becomes law, all that any submersible must do to comply with Coast Guard regulations is have adequate running lights as any ship operating in land or ocean waters.

Thus, we have four groups that are concerned with safety and certification for submersibles. But if a civilian sub is not for hire by the Navy, why should it be certified by anybody? The answer is simply that if the vehicle is to be insured it is far more likely to get a lower rate on insurance if the boat has been built and operates according to guidelines or has a certificate from ABS. The cost of insurance can run in some cases as much as half the total cost of operation. At a technical meeting, J. W. Dawson of Lloyds of London stated that Lloyds reduced the insurance rates of those vehicles which used the MTS safety guidelines.

The time is drawing near when submersibles will be regulated much as civilian and military aircraft are. The submersible industry has been extremely aware of the need for a perfect safety record and has taken on itself the job of being sure vehicles are properly built and operating safely. All major submersible builders and operators support the efforts of MTS by sending representatives to all meetings. Since the MTS guidelines are the most thorough and are generally acceptable to all builders, it is likely that they may form the basis for Congressional action. Many of the MTS regulations already have been adopted in the ABS requirements.

Once a vehicle has been properly designed for safety and after the material audit has been conducted to establish that it was built according to the design drawings, it is still necessary to stage a series of in-water tests of the vehicle. Usually, the tests are known as sea trials and they reveal whether or not the vehicle will operate as designed.

One of the most important of the tests is the series of air, surface, and submerged inclining tests. These establish the stability of the submersible which is extremely important to its safety. Also a part of sea trials are observations and documentation of descent rate, ascent rate, control characteristics, and emergency procedures. In many cases an unmanned, tethered dive is made as was done with *FNRS-2* (see p. 55). Finally, when all the bugs are eliminated and the boat is performing as designed, several shallow to medium depth dives are made—with one climaxing final deep certification dive to the full operating depth. In the case of Navy certification, someone from the Navy agency accompanies the pilots as an observer. With the completion of this dive, the vehicle is certified safe. But before a submersible may put to sea for the Navy or as a civilian sub following the MTS guidelines, there must be crew training and operational planning —both important factors in over-all safety.

It is obvious that although a submersible may be pronounced safely designed and fabricated, it's of little value if the operating crew is without adequate qualifications and training. The U.S. Navy believes crew quality to be extremely important and has issued a publication, "Procedures for Certification of Operators of Manned Non-Combatant Submersibles." These instructions spell out what a pilot must know and be able to perform before he can be certified by the Navy as being qualified to operate a Navy-certified vehicle.

Another organization, the Deep Submersible Pilots Association formed in 1966, developed a set of guidelines for submersible pilots issued in 1967. DSPA is made up of nearly all the pilots of actively operating vehicles and combines the talents and experience of many of the vehicle pilots over the past ten years. These guidelines were established to aid in the selection, training, and qualifications of pilots.

MTS includes in its "Safety and Operational Guidelines" a chapter on training, based on DSPA guidelines, which includes selection criteria, training standards, and an outline of a typical training course. A course "would include general orientation involving basic familiarization with a large number of background subjects such as ocean engineering, navigation, controls, vehicle systems, electricity, etc." This would be followed by a "ground school" including specific vehicle description, subsystems, check lists, launch and recovery systems, emergency procedures, and maintenance procedures. Next, there would be dry-run, in-water familiarization, and practice dives. A last item suggested is maintaining pilot qualifications, re-exams, and pilot logs. Several larger companies that operate submersibles have developed pilot training courses designed after the one initiated by DSPA with modifications to their own situations.

Finally, after the vehicle has been determined safe and the pilots trained, the operational procedures must adhere to some basic rules and guidelines that have been developed through experience. These operational elements are clearly outlined and discussed in a chapter of the MTS Guidelines. The most important items covered are vehicle handling equipment, support ships, support equipment, consumables, maintenance, inspection and test procedures, and operating procedures. Part of operating rules for most vehicles includes safety procedures for handling hazardous materials. Some of these are: mercury for attitude control; lithium hydroxide (used in life support); compressed air, oxygen, and nitrogen; battery electrolyte; hydraulic and other oils; caustic materials; pyrotechnics (signaling devices); heavy weights.

One of the many routine operational procedures that contributes to vehicle safety is the pre-dive checklist which helps the technicians to be sure to perform necessary preparation and show the pilots that all is in readiness before they enter the vehicle. Usually the Chief Pilot must sign an Inspection Record Sheet of some sort, accepting the vehicle. The pilots also have their own check list of predive preparations including a "walk around" outside the sub and a ten to twenty minute preparation inside. Appendix B is an example of this procedure.

Any operational plan must include a set of emergency procedures to be followed for a variety of casualty situations. In case of such events as fire, electrical short circuits, power failures, bouyancy failures, and flooding, there is an analysis of what must be done and what the suspected result will be to the vehicle and crew. These procedures should set forth operational limits of a particular vehicle for depth, time, maneuvers, bottom contact, and power consumption. Another area where preplanned procedures are essential is in the case of a communications failure, so that both the surface control and submersible crew know what to do. Many sub operators agree to surface thirty minutes after a communications failure; others, depending less on the telephone, disregard a failure and continue the dive. In this type of operation the surface ship has no cause for concern until the vehicle is overdue.

A plan has been devised by submersible operators and the U.S. Coast Guard called Mutual Assistance, Rescue, and Salvage Plan. Participating companies and Navy agencies publish important safety features of all their submersibles such as vehicle weight, size, lift points, communication frequencies, vehicle dive status, emergency control telephones, and other details. In this way, fast reaction and movement of necessary equipment to the site of a mishap may prevent a loss or fatality.

With all this talk of submersible certification and safety, maybe we should look at what it has produced, or better still, at what it has prevented. Of the estimated 5,000 to 10,000 dives made over the last few years by submersibles, very few have resulted in emergency situations; and in only one case was there a death due to a submersible accident. This is really a remarkable record, especially when compared with the early stages of development of the airplane and automobile. The concern on the part of the general submersible industry is to make sure this good practice is continued.

The one fatality occurred in 1970 and was not the result of an unsafe submersible but rather involved an unsafe operational practice. *Nekton,* the vehicle that had helped to cut the line entangling *Deep Quest* in 1969, was working with her sister sub *Nekton Beta* off Catalina Island in California. The two small 1,000-foot capability vehicles had successfully attached a lift line to a sunken cabin cruiser in about 260 feet of water. The surface ship was lifting the cruiser toward the surface when the cable parted and the cruiser plowed down, hitting the *Nekton Beta* in a one-in-a-million type of accident. The force of contact smashed a port and the vehicle filled with water, except for one

173

small pocket of trapped air. One of the occupants was able to take a gulp of air and make a free ascent to the surface. The other man was not so fortunate and drowned. Clearly, the vehicle should never have been in the water while the cruiser was being lifted, as all salvage divers would quickly point out. This accident points up the need for continued operational safety as well as vehicle design and construction safety.

In looking to the future of submersible safety, one area that is bound to expand is the private, small, wet or dry vehicle used for recreation. Of many prototypes, none has emerged yet as a significant market because of cost and other related reasons. If the recreation submersible markets should develop later on, as the snowmobile has in the last few years, there would be a definite need for safety controls and regulations. It appears that by the time a consumer recreation market would exist, there will be reasonably strict regulation of our present submersibles giving a framework for the design and operation of recreational vehicles.

# 9

# The Submersible—
# Vehicle of the Future?

In looking over only a few of the many valuable accomplishments performed by the deep submersible since 1960, it is easy enough to say that the future looks bright and promising for the development of this underwater tool. Submersible vehicles have proven themselves fit for the jobs of research, exploration, and ocean engineering.

In 1964, Dr. Fred Spiess, then Acting Director of Scripps Institution of Oceanography and longtime advocate of submarines, said: "Having to make a case for the deep submersible is like having to make a case for learning how to walk." With this conviction, he moved Scripps and many scientists ahead to participate in oceanographic research by submersible. And Richard Terry, then with North American Aviation Company, put together a thorough and up-to-date catalog of the submersible effort.

175

He called the manuscript *The Case for the Deep Submersible* and used Spiess's quotation. By 1966, when the book was published under the title *The Deep Submersible,* the case had been firmly made. Most of the people involved with undersea vehicles were quite in agreement—the case had been made—and it was a matter of moving ahead to build a fleet of submersibles and getting out to operate them.

Most of the major aerospace and technology corporations had been keeping one foot dipped in the water in case the hydrospace effort showed signs of rapid growth. The events of the mid-sixties led a number of these companies to look closely at the deep submersible. With little hesitation, many of them decided to go ahead and enter this small but promising field. This sort of corporate decision was made on the basis of enthusiasm shown by the U.S. Government—primarily the many branches of the U.S. Navy—for using submersibles in oceanography. With relatively few solid market indications, such corporations as Westinghouse Electric, Grumman Aircraft, Lockheed Aircraft, General Dynamics, and North American Rockwell, decided to spend many millions of dollars to build, equip, and operate prototype vehicles. Some saw this as a possible repeat of the glamour of the space age; others saw the design of such vehicles as a way to please stockholders; still others calculated a real return on their investment. Shortly, new vehicles began to appear—and continued to be built in steady numbers through the latter part of the sixties. The period from 1965 to 1970 saw the christening and launching of some thirty-five American vehicles. More than 75 percent of these new subs were built and financed by private industry on the speculation that there would be ample demand on the part of Government laboratories and parts of the expanding commercial market. Where, in 1965, there were no vehicles ready to operate for scientific or military missions, by 1969 there were more than enough to go around. The wonderful promise was beginning to weaken. One reason was that no one in 1965 had foreseen the enormous bite taken out of the United States economy by the escalation of the Vietnam war. The other was simply that there were too many vehicles—many that were not competitive or did not offer proper tools for accomplishing scientific tasks. The actual cost of operating submersibles with an expensive mothership was becoming apparent. Finally, the U.S. Navy had built a number of its own vehicles at a high cost that consumed all funds allotted for the program. Nearly all the tasks and missions defined by the government could be accomplished by these Navy-built vehicles—thus eliminating the need for vehicles from private industry. By 1970, it was clear that neither Congress nor the Administration was interested in providing any further funds to support the submersible industry. Consequently, the corporations which had invested well over $100 million among themselves to establish the finest fleet of submersibles in the world had to retreat and retrench. Most vehicles were stored in "moth balls" in garages and warehouses; others were sold or donated to institutions for tax purposes.

So, instead of the bright future that looked sure and just around the corner in 1965, at the time most of the submersibles were conceived and built, the industry in 1971 faced a gloomy few years where no appreciable funds would be available for chartering their vehicles.

During the last six months of 1970, at least one-half of the available submersible force—both military and private—were inactive. Yet, at that time at least five or six new vehicles were launched and christened. These new 1970 vehicles consisted of several very large and complex subs and several very small and simple ones. Both ends of the submersible scale appear to have reasons for continuing their work. A review of the latest craft shows at least three categories which are worth examining: the military mission vehicles, the experimental materials vehicles, and the research and exploration vehicles.

The loss of the *Thresher* with all 129 hands on April 10, 1963,

177

was responsible for the U.S. Navy looking very carefully at its own capability of submarine rescue and salvage. After the tragedy, the Navy formed an advisory committee, the Deep Submergence System Review Group (DSSRG). The findings of this panel showed that the Navy had no capability to rescue men trapped in a submarine below 750 feet, the maximum depth of the antiquated McCann Bell. Further, the means to detect and locate objects in deep water were extremely primitive; recovery and salvage of these objects, whether large or small, was equally impossible. The DSSRG made a number of recommendations which would remedy the Navy's shortcomings in the area of rescue and salvage. The proposed plan was endorsed in 1964 and included a series of manned deep submergence vehicles and a shallow water program called "Man in the Sea" which involved the Navy's Sea Lab experiments.

An office was set up in Washington to draw together these programs. It was called the Deep Submergence Systems Program (DSSP) and had a nucleus of Navy officers and civilians who had worked on the development of the Polaris weapons system. The first item on the DSSP agenda was the development of a workable method to rescue men from a drowned combat vehicle such as the *Thresher*. *Trieste I* had been the only vehicle able to search the wreckage site at 8,400 feet, where the *Thresher* landed. Although the operators of *Trieste* knew their vehicle was not intended for salvage work, the Navy had to see for itself what a vast improvement was needed for search and rescue, even under normal oceanic conditions. The *Deep Submergence Rescue Vehicle* (*DSRV*) was recommended for the safe rescue of a sub's crew. It was designed to locate a downed combat submarine, mate to it in water depths to 3,500 feet, and remove the cew in several trips. Originally, twelve of these 50-foot long subs were planned. They were to be stationed around the world, each one never farther than twenty-four hours from a downed submarine. The *DSRV* was designed to fit into a C-141 cargo aircraft or the larger C-5A for delivery to the airport nearest Auxiliary Submarine Rescue, the Navy's catamaran submarine rescue ship. Or, if the location of the downed submarine was too far from port, the *DSRV* could be carried piggyback on one of the fleet nuclear submarines which, at 15 knots, is faster than a surface ship.

The competitive bidding between the major corporations on the *DSRV* was active since it appeared to be a fat plum for the builder. The initial contract went to Lockheed, who believed they might build up to twelve vehicles. However, as costs began to mount in 1966, and the Navy found that a *DSRV* could be designed to rescue up to twenty-four persons at a time instead of only sixteen, the number of vehicles was reduced to a total of just six. The vehicle which the Navy asked Lockheed to build was an overwhelmingly complex machine involving highly sensitive and elaborate instrumentation never before attempted in a submersible. Instead of the relatively simple controls employed in *Deepstar, Alvin,* or *Aluminaut* the Navy and DSSP believed that to maneuver the 75,000-pound vehicle to its target required an inertial guidance system mated with a flight computer similar to those used in a Boeing 707 aircraft. A crew of three was housed in a forward sphere that was packed with instrument readouts, TV monitors, banks of warning lights and status indicators, and a computer command board. The design and fabrication of the instrumentation and control, called the ICAD, for Integrated Control And Display, was carried out by the MIT Instrumentation Laboratory, Boston, Massachusetts. The vehicle's pressure hull is made of three intersecting spheres. The after two spheres are the rescue chambers for the twenty-four men. The center sphere has a large flared skirt on the bottom which is used to mate with the hatch of a submarine. *DSRV* can be maneuvered to tilt at angles of 45° in roll or pitch, if necessary, to permit mating with a submarine lying on its side.

One of the technical difficulties encountered was that little of the "off-the-shelf" equipment was available or could stand the

179

rigors of deep sea military specifications. So the costs continued to soar in the building program. After the *Scorpion* disaster, in which the nuclear submarine sank in 10,000 feet of water off the Azores in 1968, DSSP concluded that future sinkings were statistically more likely in deep water than shallow, making rescue less possible than originally thought and it is likely that a combatant submarine would probably be crushed at about 3,500 feet. Finally, when the previously-allocated funds were devoured by the Vietnam war effort, it was decided that money was available to build and operate only two *DSRVs*.

After about five years and roughly $50 million, the first vehicle, *DSRV-1*, was launched on January 29, 1970, in a ceremony at San Diego by the Lockheed Missile and Space Company. The first boat was certified for 3,500 feet initially but following additional testing will be rated for 5,000 feet along with the *DSRV-2*.

Sea trials, in which all bugs are supposed to be worked out, continued during 1970 for *DSRV-1*. A final deep dive to maximum operating depth was performed in December. Following this and some mating tests the Navy took over command in May 1971.

The *DSRVs* can be remarkably valuable rescue vehicles if the complexities of the ship do not interfere with maintenance and operation at sea. In the first half of 1971, neither *DSRV* had a permanent surface support ship, since the program to build the ASRs also encountered major financing problems. However two of the Navy's military submarines were equipped to carry *DSRV* piggyback if necessary. The *DSRVs* are based in San Diego and operated by the Sub Development Group.

The companion to *DSRV* is the *DSSV* or Search Vehicle that was recommended at the same time as the *DSRV* to give the Navy a vehicle capable of searching for objects lying at 20,000-foot depths. Twenty-thousand feet has been the depth selected as representing at least 98 percent of the area bottom, which leaves out that one or two percent of extremely deep trenches such as *Trieste* explored on Project Nekton.

The *DSSV* had the basic requirements to carry a crew of four for up to thirty-six hours at 4 to 5 knots on a search mission that would employ a sophisticated side-looking sonar to locate objects as small as six feet in diameter. For 20,000-foot missions, *DSSV* must have a strong, thick-walled pressure cabin and would require considerable flotation to offset this weight. Further, the time and relatively high speed requirements meant enormous volumes of batteries for propulsion. The "SV," as it was known in contrast with the "*RV*," would cost considerably more to build because of more instrumentation for searching and the need for deeper and longer missions.

In 1967, the Navy entered into a preliminary design contract with two major corporations, Westinghouse and Lockheed, both of whom had previous submersible experience. Westinghouse had operated the *Deepstar* and *Diving Saucer,* and Lockheed had recently completed its own vehicle, *Deep Quest,* and was in the midst of the *DSRV* program which had many similarities to the *DSSV*. The design competition, the second phase of the vehicle construction, was won by Lockheed in 1968. The cost estimates showed that between $60 and $100 million would be needed to complete the 50-foot long boat. During 1969, it was obvious that such an undertaking could not be accomplished by the Navy on its shrinking budget. In January, 1970, the *DSSV* program was cancelled, after Westinghouse and Lockheed had each invested at least $2 million of their own monies in the design development program. Thus the Navy's deep search capability, although improved by towed, unmanned sonars, is still seriously limited.

Another new approach to submersibles has been the development of a small, nuclear research sub for relatively deep diving and long-term missions. Funds for the *Nuclear Research Vehicle* (*NR-1*) were appropriated about 1965, and work began shortly thereafter at Electric Boat Yards in Groton, Connecticut. The program was under the strong,sheltering wing of Admiral Hyman

181

Rickover, father of the nuclear submarine program. He has since put much of the details of the 137-foot long, 740-ton ship under tight security. The *NR-1* was launched in 1969 and was operated for at least a year. Her mission is primarily search with a suspected depth limit of 3,000 feet. Secondary missions will be for exploratory and oceanographic projects. She carries instrumentation of complexity equal to *DSRV*. A pilot and copilot are seated in a wraparound console. Each man has two control sticks: one controls roll, yaw, pitch; the other, surge, sway and heave. These are the six basic motions of a submersible. *NR-1* has both an analog computer for basic ship control computations and two microelectronic digital computers to aid in navigational calculations using information received from doppler sonar and underwater beacons and transponders.

*NR-1* carries a basic crew of five and two scientist/observers for a thirty-day mission. The vehicle has a 2,000-pound payload which includes still, movie, and television cameras, viewpoints, a grab sampler, a manipulator, and a fresh water still.

The cost of *NR-1* was originally estimated to be around $25 million, a large part of which was for the small pressurized water-nuclear reactor that supplied the power. Like the *DSRV* and its ill-fated cousin *DSSV*, the bills for *NR-1* began to mount rapidly as Admiral Rickover realized that few of the existing sensors that the Navy thought were adequate could endure the extended mission periods. External lights were especially vulnerable. Since all other submersibles were usually out of the water most of the time, it was easy enough to change the light bulb. But *NR-1* demanded lifetimes for exterior lights that exceeded the normal 500 hours expected of incandescent bulbs or even some of the advanced vapor lights. Therefore new types of light elements had to be developed.

By launching time, the *NR-1* cost had passed the $90 million mark. Under the strong hand of Admiral Rickover, it is likely that little was compromised in the proper equipping of the boat and it survived the cuts that sank the *DSSV*. As to the operating depth of *NR-1*, while the Navy has released no information on an exact depth, it has stated that *NR-1* has gone deeper than any other nuclear boat. Various published guesses put it in the vicinity of 3,000 feet. This is reasonably deep, considering that the pressure hull is nearly 125 feet long and 8 feet in diameter.

With the nuclear power, long range, and sophisticated instrumentation, the *NR-1* may be the most outstanding vehicle afloat. But because of its military missions, we may never know any more details about it.

Another boat in the category of both military and oceanographic research is the *Dolphin*, AGSS 555, 152 feet in length and 900 tons displacement. *Dolphin* was designed as a conventional submarine with diesel-electric propulsion and short-term duration, but is capable of deep depths for experimental research in sonar and Anti-Submarine Warfare (ASW). The constant diameter hull permits her to operate deeper than *NR-1*. Her size allows her to be an acoustic platform, to carry large transducers and to simulate a submarine target. Unlike all her sister submersibles, except for *NR-1*, she is independent of a surface support ship and is thus defined a submarine. Of course, with only batteries for underwater propulsion, she must surface frequently to charge them as did the World War II submarines.

Construction on *Dolphin* was started in 1962 at Portsmouth Navy Yard, New Hampshire. *Dolphin* was reportedly under construction longer than any other U.S. submarine, due to design problems and the lack of funds for development. *Dolphin*, along with the *DSRVs* and *Trieste II*, is stationed in San Diego where she is operated by the Navy's Submarine Development Group One (SUBDEVGRU ONE).

*Dolphin* has a crew complement of fifteen to twenty-three, and can accommodate seven scientists. She was commissioned in 1968 and after outfitting and trials she sailed from New England to San Diego arriving for duty in November 1970. Although she hardly has the comforts or space of a nuclear boat, her great depth capability makes her an important vehicle for the Navy.

The submersible era began with *Trieste I* as a research submersible. It was later modified in *Trieste II* for the completion of the *Thresher* search. The latest configuration for *Trieste* is considerably different from the one in 1965. It now has a totally streamlined shape which allows improved towing speeds. It was used for the *Scorpion* search and bottom investigation and photographed the wreckage of *Scorpion* at 10,000 feet. Among the many sensors under development for deep submergence work is a unique television camera called El Tortuga. This camera was used by *Trieste* during the 1969 search for *Scorpion*. A conventional TV camera was housed in a syntactic foam case that made it neutrally buoyant. A set of small water pumps with ducted jets were incorporated along with a light. This contraption, controlled from the submersible, used the water jets to fly the camera which pulled along its umbilical cable. *Trieste II* used two of these cameras to explore the wreckage of the downed submarine. These cameras were able to get into inaccessible nooks impossible to the larger vehicle.

As indicated in Chapter Five, experimental work has been continuing on new materials for pressure hull construction. All those who have spent any time diving in a submersible will agree that anything that can improve visibility is worth doing. For this reason, a group at Navy Underwater Center have spent a number of years working toward the use of transparent hull materials. Of the possible candidate materials, glass offers a real promise. The first steps toward high visibility were taken in the use of acrylic plastic, which had been in use for view ports for a long time. The first successful attempt at an acrylic sphere for people was made with *Nemo*. This two-man, 6-foot sphere uses a heavy wall of clear plastic which gives full vision in all directions. It is cleared for 600-foot depths and in 1970 made a dive that deep in the Bahamas. Acrylic plastic appears to be practical to a maxi-

# The Submersible—Vehicle of the Future?

mum of 2,000 feet, at which depth the wall must be at least 4 inches thick, as in the *Johnson-Sea-Link*. Beyond 4 inches, the weight due to thickness becomes so great that acrylic plastic is impractical. Therefore, subsequent vehicles such as *Nemo* will be valuable in the relatively shallow water where they offer the greatest advantages. From experience gained with this vehicle will come recreational vehicles made entirely of clear acrylic. A prototype of this sort which can dive with two passengers to a maximum of 100 feet was under-going trial in San Diego in 1970.

For deep depth and high visibility, structural glass appears to be winning support, although slowly. Experimental work underway at NUC at San Diego, directed by Will Forman, involves the use of a glass hemispherical head in a two-man vehicle called *Deep View*. This vehicle is considered experimental in nature and is intended to prove the feasibility of glass for high visibility and high strength.

*Deep View* may ultimately go to 2,000 feet, after being readied for dives to shallower depths. It will work by steps into deeper water. The glass forward end is held in place by a titanium retaining ring which connects it to the steel cylindrical hull.

The ability to see in any direction and all directions simultaneously is probably one of the most important advances in submersible technology.

One of the most ambitious of the research submersibles projects was put into the design stage in 1966 by Westinghouse. This was *Deepstar 20,000*, the third member of the *Deepstar* family conceived several years earlier. *Deepstar 20,000*, carrying three men, was intended for 20,000-foot depths and has many of the basic features found in *Deepstars 4000* and *2000*. The first mission proposed for it included deep diving for acoustic, geophysical, and physical oceanography. Possible charter parties were mainly the two Navy labs, U.S. Navy Underwater Sound Lab (USN/USL), and U.S. Navy Underwater Center (USNUC). Both wanted to do research from a submersible at depths of

185

20,000 feet.

The progress of *Deepstar 20,000* was slow because the project was financed by Westinghouse itself. By 1969, work had begun on the single spherical hull. Experiments had been conducted earlier on various types of steel that could offer the highest strength-to-weight ratio. One was a new steel called HP 9-4-20. Another was the more conventional HY-140 type. After full analysis by Westinghouse it was decided that the HY-140 steel was better suited to the *Deepstar* pressure hull requirements. During 1969, Westinghouse had begun to increase the capabilities of *Deepstar 20,000* in the hope that it would be more marketable for leasing. Since Westinghouse had also worked on the preliminary design of the Navy's *DSSV*, some of the innovations planned for that vehicle were incorporated into *Deepstar*. One of these was the unique Varivec propulsion system. In this system one single motor drives the stern propeller that controls forward and astern thrust as well as vertical or side thrust.

*Deepstar* was reconfigured to accomplish a longer bottom search using the highly sophisticated side-looking sonar designed and built by Westinghouse. All steps of the construction were documented to be able to meet the Navy's rigorous certification requirements. As a result of the modifications, the vehicle's overall weight rose rapidly. Instead of the original planned air-weight of less than 40,000 pounds, the vehicle more than doubled to 85,000 pounds with an over-all length of 36 feet. This was due to the HY-140 steel sphere, measuring 7 feet in inside diameter and weighing 18,700 pounds, instead of the lighter hull planned using 9-4-20 steel and syntactic foam flotation using 42 lb./cu. ft. foam instead of 36 lb./cu. ft. Some 20 tons of this heavier material was required in the design. The exostructure or outer framework and skin, which was free-flooding, was to be made of epoxy fiberglas. Work on all the subsystems continued through 1970 with hope for a launch in 1971. As far as being able to drive to the bottom of nearly all oceans—or at least 98 percent of them—

# The Submersible—Vehicle of the Future?

*Deepstar 20,000* looked promising. It appeared that the completed boat would cost Westinghouse between $5 and $10 million. Early in 1970, there appeared to be a solid customer in the Navy, until the *DSSV* submersible was cancelled along with the parent office DSSP, which was closed down mid-year due to lack of additional funds. Finally Westinghouse saw all too clearly that there would be no federal monies in the next few years for a submersible like *Deepstar* with a diving capacity of 20,000 feet. The *Deepstar 20,000* program was halted in the fall of 1970, and the still-incomplete and yet-to-be-assembled pieces were crated to join the smaller sister *Deepstar 4000* in moth balls.

Other research submersibles have been more fortunate. Two of these had been on the drawing board since 1965 and were likely to become the Bobbsey Twins of the Navy scientific vehicles. Originally, these vehicles had been given the name of the AUTEC boats, since one or both were going to be working in the Atlantic Underwater Test and Evaluation Center (AUTEC) in Tongue of the Ocean, Bahama Islands, where the Navy has a test range. The AUTEC boats, *Autec I* and *Alvin II*, with some modifications, were actually copies of *Alvin*, owned by the Navy. The way in which these two vehicles came into being is interesting since it was not planned but rather, just happened.

*Alvin* had been built with two spare pressure hulls fabricated by Lukens Steel Corporation, since those in charge realized that the cost of an additional hull is relatively low. The Woods Hole Oceanographic and Navy officials had thought it wise to determine the ultimate strength of the *Alvin* pressure hull by pressure testing one to the point of destruction, rather than guessing at what point failure might occur. So, during the construction, of *Alvin*, one of the hulls was sent to Southwest Research Institute in San Antonio, Texas, known for its experience with pressure vessels. The hull was placed in the largest pressure test chamber available and was subjected to a pressure equivalent to a 9,000-foot depth. But, instead of the expected failure of the

187

hull by bending or deflection, the test chamber burst! The designers and builders had exceeded the requirements, and everyone (except the owners of the chamber) was happy because it meant that *Alvin's* pressure sphere, normally rated for 6,000 feet, had a safety factor of nearly 2 to 1. Further, since the *Alvin* was a Navy project sponsored by ONR, the Navy now had not one but two spare hulls just like *Alvin's*. It was only natural to build two more *Alvin*-type vehicles, for here were the hulls already built and by just adding some equipment, the Navy would have a flotilla of three vehicles nearly alike.

And this is exactly what the Navy did; in 1965 bids were sought to construct two *Alvin*-type vehicles. The low bidder was the Electric Boat Division of General Dynamics which bid to build the two for slightly over $3 million. The vehicles were scheduled for delivery in 1967, but, as is typical with developmental projects, there were innumerable delays, changes and revisions to the design, as well as the relentless pressure for more money. One vehicle, *Alvin II* was to go to Woods Hole (WHOI); the other to go to a group at AUTEC. But 1967 came and went and progress was slow. The vehicles were 26 feet, slightly larger than *Alvin's* 23 feet, and, where *Alvin* weighed 16 tons in air, the AUTEC boats tipped the scales at 25 tons each. They each were to carry three men to 6,500 feet, but many of the vehicle systems were modified for greater safety by the Navy and thus became more complex and consequently were increasingly heavier.

In rushing into the construction of these boats, several important points were overlooked that have compromised the value of these submersibles. In late 1964, when the Navy decided to proceed with *Autec* boats, *Alvin* was still in sea trials and WHOI was just beginning to learn how to use her. The interior layout was partially modified for better placement of the pilot and observer. During initial operations in 1965, they found the layout of the human accommodations was still not right and so, in 1966, a second, more suitable crew arrangement was adopted. It was long after the advertised bid for the AUTEC boats that WHOI found the best way to accommodate to the poor placement of view ports in the hull, but no amount of rearranging could overcome the fact that the ports did not have overlapping views. Thus, the pilot and observer never were sure that they were both looking at the same scene. In a submersible used for scientific purposes where much depends on visibility, this was a serious shortcoming. Yet even with this less-than-ideal port arrangement, the Navy had decided to use the two spare hulls.

Of the many valuable modifications made to *Alvin* through learning and trial and error over several years of operation, little could be incorporated into the new boats because the design had to be "frozen" at Electric Boat at an early stage, if the contractual delivery date and cost were to be met. In fact, the first vehicle, *Alvin*, was more up-to-date and better able to do her scientific mission than the newer boats still on the assembly line.

The other problem that faces these new vehicles is the lack of a surface craft designed to handle 25 tons of weight. The basic design did not allow for a single lift point as with the *Deepstars*, *Beaver*, and others. The AUTEC boats could only be picked up from beneath on their landing skids and required far heavier and more elaborate lift elevators. This severely limited the type of surface mother ship. By 1969, the Navy had had to spend more money on the project than originally allotted and there was none left for an adequate mother ship.

On December 11, 1968, the vehicles were christened at Groton, Connecticut. They were given the name *Sea Cliff* for Sea Cliff, Long Island, and *Turtle*, for Turtletown, Tennessee. Finally after some finishing touches, the vehicles were put through a long drawn-out set of sea trials by the Navy in the Bahamas during the summer of 1970. Still, no mother ship had been obtained. This made operations extremely difficult and slow. To get to deep water during trials, the vehicle had to be towed out from port at "all ahead, dead slow" since the large foreward area was hardly a hydrodynamic shape. Worse still was the launch and recovery procedure. A crane was used, but since a four-point strong back or support bar had to be used also, a diver had to hook on the shackles. This could only be done safely in not more than ½-foot waves, a very limited condition anywhere at sea. This points out the utmost importance of a mother ship being part of the initial design of every submersible.

The final episode in the long saga of *Sea Cliff* and *Turtle* was slightly more promising. Instead of working in the AUTEC range with WHOI, where operating funds were limited, both submersibles were assigned to the Submarine Development Group One, making a total of six submarines owned and operated by the U.S. Navy in San Diego.

So, while submersibles such as *Sea Cliff* and *Turtle, DSRV-1* and *2, Dolphin* and *Trieste II* were just getting into gear to begin an active career of operations, a larger number of civilian-owned-and-operated craft had been laid up—some perhaps never to return to the sea. One of the most disastrous aspects of a cutback such as the one that started in 1970 is the breaking up of operating and design teams built up over a six- or seven-year period. The loss of men experienced as operators and engineers is unfortunate because once the submersible business is active again —and there is no doubt that it will be—the time for retraining and relearning will have to be provided all over again.

Before looking at where the submersible is presently going and what its future employment may be, we should not overlook the activities of several countries where many of the original efforts began over fifteen years ago and interesting developments are now underway.

Some of the most successful and significant submersible vehicles have been designed and built in France. Among the operating ones are *Diving Saucer* (*SP-300*), *Deepstar 4000,* the two *Sea Fleas* (*SP-500*), *Soucoupe Plongeante* (*SP-3000*), and *Archi-*

# The Submersible—Vehicle of the Future?

*mede.* In 1970 several others were under construction or in preliminary design. Although *Archimede* has been diving since 1961 and is probably the only full-depth submersible in existence, she has had little publicity or acclaim. *Archimede* was the successor to the *FNRS-3* and was designed by the French Navy to dive to the deepest part of the Pacific Ocean. Of course, as we know, the Project Nekton dive of *Trieste I* beat *Archimede* by nearly two years. *Archimede* is nearly 75 feet in length and weighs 200 tons. This large boat dived six times in 1962 in the Kurile Trench off Japan to about 28,000 feet. The deepest dive made was 31,-000 feet, although *Archimede* can safely dive to 36,000 feet. Later, in 1964, she made a detailed study of the Puerto Rican Trench diving to 25,000 feet for the U.S. Navy (ONR) and Centre Scientifique Recherches France. *Archimede* is equipped with cameras and a rock corer. The French Navy has used her to search for the military submarine *Euridice* in the Mediterranean. While greatly unheralded and with relatively few scientific papers resulting from the dives, *Archimede* is still the sole research submersible capable of extreme depths.

Captain Cousteau is undoubtedly the most active force in submersible activity in France. Following the delivery of *Deepstar 4000* to Westinghouse in the United States, his design team began work on a one-man vehicle that could carry movie cameras and lights to 2,000 feet. Two of the *Puce de Mer* or *Sea Fleas* (*SP-500*), also better known as the mini-subs, were constructed in 1967. One of the missions of these boats was to help in making movie sequences for the Cousteau television film series. The two vehicles have been used together as scuba divers use the "buddy system." The mini-subs are 10 feet long, weigh about 2 tons each, and are carried aboard *Calypso.* They have been used for filming and exploration off California, Mexico, and in the Carribean.

The newest in the Cousteau Diving Saucer family is the *Soucoupe Plongeante 3000* (*SP-3000*) which began sea trials in 1970.

# Diving for Science

The design is close to that of *Deepstar 4000,* using many similar components. The pressure hull came from the first attempt to build *Deepstar 4000,* when the Vasco-Jet 90 steel did not meet Westinghouse and U.S. Navy certification requirements. Additional test data gathered since 1964 showed that the pressure hull could be rated as nearly 10,000 feet. *SP-3000* is really an extension to 10,000 feet of *Deepstar 4000,* promising to be useful to the French scientific research program. She was scheduled to dive in the Mediterranean in 1971 and in the mid-Atlantic Ridge in 1972 and 1973.

One of the most exciting ventures into submersible technology will be the completion of *Argyronete,* named for the water spider that creates tiny nests of air bubbles underwater. The principal feature of this 80-foot submarine is to combine a saturation diving (mixed breathing gas) capability and a submarine supporting six crew and four divers for up to ten days. *Argyronete* will be independent of the surface with a range of 400 miles. On the surface, she has diesel power and for this reason is considered a full-fledged submarine. She is being built for Centre National Pour l'Exploitation des Oceans (CNEXO), the French oceanographic agency, and Institute of French Petroleum (IFP). Her mission will be to take four divers to depths as great as 2,000 feet to work on petroleum production projects. *Argyronete* is to be operational by 1973 unless the general slowdown of oceanographic investigations delays the schedule.

After World War II, Japanese scientists began to explore the shallow waters around their home islands to determine ways to improve their fisheries and mineral resources. The first vehicle, *Kuroshio,* a tethered chamber, was used by physical and geologic oceanographers to explore the continental shelf to 650 feet beginning in 1951. *Kuroshio II,* built in 1960, was a four-man tethered submersible, an improvement of the chamber. It was engaged in the same sort of work, especially in fisheries investigations. This vehicle has been important in giving Japanese

192

engineers a trial horse around which to design the improved vehicles that have followed.

The first of the "free" submersibles was *Yomiuri*. In 1964, it dived to 1,000 feet with a crew of six men. Since then, it has made over 450 working dives for scientists in Japan, primarily studying sea-floor characteristics and geologic processes. *Yomiuri* is owned by Yomiuri Shinbun, one of the leading newspapers in Japan.

Another step in Japan's progress down into the sea came in 1968 when a much improved sub was completed and *Shinkai* made her maiden dive to 2,000 feet. The 50-foot, four-man vehicle, was built by Kawasaki Industries and began operation for scientists in 1970. *Shinkai* uses a two-sphere pressure hull with a joining tunnel. The after sphere houses equipment and the forward one personnel.

Design work began in 1970 on a 20,000 foot submersible of 40 tons at Kawasaki and Mitsubishi Industries which is planned to be operational by 1973.

The main underwater research work in Russia has been carried out by the converted fleet submarine from World War II called *Severyanka* which only operates in shallow depths. A number of tethered chambers have been employed for many years—several prior to World War II. Also, a number of free-flooding vehicles have been used for investigations of shallow water fisheries.

The principal submersible project has been *Sever 2*. Translations of various articles and reports by visitors to the USSR relate the slow progress on this research vehicle. At one time, the 33-foot sub designed for 6,000 foot depths was said to resemble *Alvin* in appearance and capability, but the latest article on *Sever 2* describes the 28-ton vehicle as looking more like *Aluminaut*. She has been under construction for six years and has been reportedly tested unmanned in 1970 to a depth of 6,662 feet. *Sever 2* is under the control of the fisheries research group which

193

has little interest or money for deep diving but has spent most of its effort on shallow water research. However, the Institute of Oceanology, part of the Soviet Academy of Science, is more interested in deep submersible work and has the money to carry out its interest. The Russian group purchased *Pisces IV* from the Canadian company, International Hydrodynamics, in March, 1971 for $2 million. *Pisces IV* has a diving capability of 6,600 feet for three men. It carries two manipulators as well as a special package of oceanographic instruments to be used on scientific dives by the Russians.

The USSR attemped to purchase *Star III*, the 2,000-foot, HY-100 steel hull sub from Electric Boat Company in 1968. However, the United States Government refused to allow the sale.

The first and only real submersible in Britain is *Surv*, launched in 1968 into the English Channel. The 10-foot, 5-ton vehicle was designed to carry two observers to 600 feet. *Surv* was built by Lintott Engineering, Ltd. for use by The National Institute of Oceanography in geologic sampling and photography. Initial trials were discouraging, as is the case many times until all the bugs are worked out. *Surv* experienced the humiliation of blowing a static inverter for its AC propulsion, a failure familiar to those who use AC systems. Its basic design resembles *Deep Jeep* and *DOWB*.

Another effort in Britain that appears to have failed is the Cammel-Laird Seabed *Crawler*. The *Crawler* is a four-wheeled vehicle that began shallow water mining survey operations along the continental shelf in 1970.

One company has been very active in building submersibles in Canada mostly for the American market. International Hydrodynamics (HYCO) originally built the *Pisces I* in 1965. This 1,800-foot-depth sub has made over 500 dives in a large variety of assignments. Three other vehicles of similar design followed with capabilities ranging to 3,000 feet. Two of these are scheduled to be used by the parent corporation Vickers of Great Britain

194

# The Submersible—Vehicle of the Future?

aboard their specially designed mother ship.

The most interesting sub of the HYCO series, and certainly the most promising, is the *CLS-1,* launched in late 1970. This was the first submersible to be purchased by the Canadian government. Its main feature is the lock-out capability, which allows divers to swim out and return on scientific and engineering tasks.

From this very brief review of submersible activities in other countries, it is obvious that there is an interest and involvement in many ways equal to that of the United States—if not in quantity, certainly in quality. None appears to have spent the enormous amounts of money on relatively sophisticated vehicles such as those sponsored by the U.S. Navy. The vehicles designed for scientific research abroad are about on a level with those built in this country, resulting from the cross-fertilization of vehicles like the *Deepstars* and *Trieste I* and their respective lineage.

As the pioneers of the aircraft industry have demonstrated through the growth of an infant business in the 1920s to its enormous proportions today, it is nearly impossible to predict what may come thirty to fifty years from now in the undersea craft industry. The negative forecasts of the 1920s and the firm conviction that it was absurd and impractical to create a passenger-carrying airplane industry are a good example. This is not to say that it is likely that a major transportation industry will result from our initial submersible efforts. But it would be foolhardy in view of the limited forecasts of past decades to say that it will not happen.

The research submersible used for deep-diving scientific and military missions has made its case and justified its existence. But yet it has not arrived, nor insured its future at this uncertain point. In a time of short funds for research and development, the submersible has been the first to suffer because of the relative high cost of construction and operation, as compared with conventional surface techniques. Such lean times may be viewed in both a good and poor light. By forcing operating subs to economize and be more efficient, we can hope the resulting successful boats will be improvements on what we're using today. This means the areas such as surface handling and mother ships, which constitute a large part of the cost, may be eliminated or vastly changed. As we decide what types of vehicle are best suited for particular kinds of diving research and specific missions, there will be standardization of designs and components that will surely reduce costs and increase reliability.

The promise for a privately-owned or corporately-owned submersible is not encouraging as a business venture in the early 1970s. It appears that only about one in five were in use in 1971. Most of these were the smaller, less-expensive vessels that could be used for short periods at less cost than when a large mother ship was involved—or where the sub was already part of a complete system such as *Deep Quest* with her mother ship.

The present trend is clear. Many of the vehicles in the United States will have to be mothballed for several years. It is quite probable that a lot of these stored submersibles will not be reactivated when funds and interest are restored. More likely, newer vehicles will be designed and built because new technologies will certainly provide marked improvements. It seems a shame that some submersibles will become museum pieces only a short while after sea trials. But survival goes to the fittest, and this means that the sub more economical to operate is likely to succeed.

One aspect of the submersible that has not been developed nor explored to any extent is the design of a recreational vehicle that could be used for shallow water exploration and enjoyment. Although over fifty different types of wet or ambient-pressure boats have been built, mostly for military purposes, very few dry boats of the same simple shallow-water design have been built. One of the few is the *VAST Mark III*, a 250-foot depth, one-man boat. A similar two-man boat built for a mass market at the price of an average sports car could be enormously successful in the recreation market. Then again, if such a vehicle were the counterpart to the snowmobile, it could be the undoing of the fragile underwater environment.

In a nation where within one generation we progressed from a Buck Rogers fantasy to men actually walking on the moon, it seems hard to believe that we won't see similar fantastic accomplishments with undersea vehicles. There is little doubt in the minds of the scientists and engineers who have used our latest submersibles that, properly used, this vehicle promises to be of enormous value in the future.

# Appendices

## A.  Vehicle Statistics

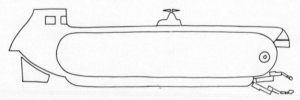

**ALUMINAUT**     LAUNCHED  1964
BUILDER   General Dynamics/Electric Boat Div.
OWNER/OPERATOR   Reynolds Submarine Services, Miami, Fla.

| | | | |
|---|---|---|---|
| NUMBER OF DIVES | 200 | LIFE SUPPORT | 450 hrs. max. |
| TOTAL DIVE HOURS | 2,000 | PILOTS 2 | OBSERVERS 5 |
| MAX. OPERATING DEPTH | 15,000 ft.* | VIEWPORTS 4 | |
| MAX. DEPTH ACHIEVED | 6,250 ft. | MANIPULATORS 2 | electric—general |
| COLLAPSE DEPTH | 22,500 ft. | | purpose |
| HULL SIZE | 96 in. | PROPULSION (2) 5-hp stern; | |
| LENGTH 51 ft. BEAM | 15.5 ft. | (1) vertical propeller | |
| HEIGHT 14 ft. WEIGHT | 146,000 lbs. | POWER silver zinc battery 64 kwh. | |
| MATERIAL | aluminum | CRUISE SPEED | 2.5 knots for 32 hrs. |
| PAYLOAD | 6,000 lbs. | MAX. SPEED | 3.5 knots for 12 hrs. |

MAJOR EQUIPMENT   Gyrocompass, UQC surface radio, altitude/depth sonar,
CTFM sonar, side-looking sonar, still and movie
cameras, pinger
ESTIMATED COST   Not available
STATUS   In semipermanent storage in Jacksonville, Fla.

* Design depth with modified hemi-heads; present capability, 8–10,000 ft.

# Appendices

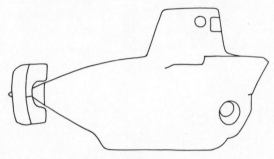

**ALVIN**     LAUNCHED   June 1964
BUILDER   General Mills/Litton Industries, Minneapolis, Minn.
OWNER/OPERATOR   U.S. Navy/ONR/Woods Hole Oceanographic Institution, Woods Hole, Mass.

| | | | |
|---|---|---|---|
| NUMBER OF DIVES | 307 | PILOTS 1       OBSERVERS 2 | |
| TOTAL DIVE HOURS | 1,200 | VIEWPORTS 4 forward/down and | |
| MAX. OPERATING DEPTH | 6,000 ft. | side | |
| MAX. DEPTH ACHIEVED | 7,500 ft. | MANIPULATORS 1 electric — with | |
| COLLAPSE DEPTH | 16,100 ft. | tools/corer nets | |
| HULL SIZE | 79 in. | PROPULSION (1) 7-hp stern hydraulic; | |
| LENGTH 23 ft.   BEAM | 8.5 ft. | (2) 2-hp side | |
| HEIGHT 13 ft.   WEIGHT | 32,015 lbs. | POWER lead acid battery 40 kwh. | |
| MATERIAL | HY-100 steel | CRUISE SPEED 1 knot for 8 hrs. | |
| PAYLOAD | 1,200 lbs. | MAX. SPEED 2 knots for 2 hrs. | |
| LIFE SUPPORT | 80 hrs. max. | | |

MAJOR EQUIPMENT   Gyrocompass, CTFM sonar, precision depth system, TV, UQC, tracking beacon, still cameras and lights, current meter, STD with digitizer, radio
ESTIMATED COST   Initial contract $575,000
STATUS   Began diving in June 1971 after complete overhaul. Titanium hull to be installed in mid-1972 for 12,000-foot depths.

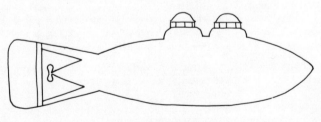

**AMERSUB 300**

# Appendices

**AMERSUB 300**  LAUNCHED  1961
BUILDER  American Submarine Co.
OWNER/OPERATOR  Underwater Inc., Mako Products Div.

| | | | |
|---|---|---|---|
| NUMBER OF DIVES | n. a. | PAYLOAD | 450 lbs. |
| TOTAL DIVE HOURS | n. a. | LIFE SUPPORT | 16 hrs. max. |
| MAX. OPERATING DEPTH | 300 ft. | PILOTS 1 | OBSERVERS 1 |
| MAX. DEPTH ACHIEVED | 300 ft. | VIEWPORTS | 360° in conning tower |
| COLLAPSE DEPTH | 1,000 ft. | MANIPULATORS | none |
| HULL SIZE | 40 in. | PROPULSION | (1) 3-hp DC motor |
| LENGTH 12 ft. BEAM | 4.4 ft. | POWER | lead acid battery |
| HEIGHT 4.9 ft. WEIGHT | 2,000 lbs. | CRUISE SPEED | 1 knot for 8 hrs. |
| MATERIAL | steel | MAX. SPEED | 4 knots for 3 hrs. |

MAJOR EQUIPMENT  Unknown
ESTIMATED COST  $9,000
STATUS  Unknown

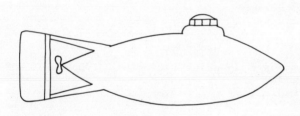

**AMERSUB 600**  LAUNCHED  1963
BUILDER  American Submarine Co.
OWNER/OPERATOR  Underwater Inc., Mako Products Div.

| | | | |
|---|---|---|---|
| NUMBER OF DIVES | n. a. | PAYLOAD | 700 lbs. |
| TOTAL DIVE HOURS | n. a. | LIFE SUPPORT | 16 hrs. max. |
| MAX. OPERATING DEPTH | 600 ft. | PILOTS 1 | OBSERVERS 1 |
| MAX. DEPTH ACHIEVED | 600 ft. | VIEWPORTS | 360° in conning tower |
| COLLAPSE DEPTH | 1,500 ft. | MANIPULATORS | none |
| HULL SIZE | 37 in. | PROPULSION | (1) 3.5-hp DC motor |
| LENGTH 13 ft. BEAM | 44 ft. | POWER | lead acid battery |
| HEIGHT 4.9 ft. WEIGHT | 3,500 lbs. | CRUISE SPEED | 1 knot for 10 hrs. |
| MATERIAL | steel | MAX. SPEED | 3 knots for 6 hrs. |

MAJOR EQUIPMENT  Unknown
ESTIMATED COST  $12,000
STATUS  Unknown

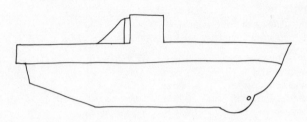

**ARCHIMEDE**   LAUNCHED   1961
BUILDER   French Navy
OWNER/OPERATOR   French Navy, Toulon, France—Groupe des Bathyscaphes

| | | | |
|---|---|---|---|
| NUMBER OF DIVES | 181 | LIFE SUPPORT | 100 hrs. max. |
| TOTAL DIVE HOURS | 1,450 | PILOTS 2 | OBSERVERS 1 |
| MAX. OPERATING DEPTH | 36,000 ft. | VIEWPORTS 3 | |
| MAX. DEPTH ACHIEVED | 31,600 ft. | MANIPULATORS 2—general purpose | |
| COLLAPSE DEPTH | 100,000 | and rock drill | |
| HULL SIZE | 84 in. | PROPULSION (1) 20-hp DC stern; | |
| LENGTH 69 ft. BEAM | 17 ft. | 2 thrusters, 5 hp | |
| HEIGHT 26 ft. WEIGHT | 200 tons | POWER alkaline battery 100 kwh. | |
| MATERIAL | steel | CRUISE SPEED 0.5 knots for 10 hrs. | |
| PAYLOAD | 6,000 lbs. | MAX. SPEED 2.5 knots for 3 hrs. | |

MAJOR EQUIPMENT   TV camera, 3 still cameras, 6 strobes, rock drill
ESTIMATED COST   n. a.
STATUS   Made series of scientific dives in 1971 in the Mediterranean.

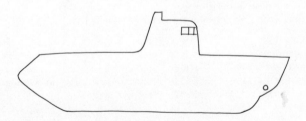

**ARGYRONETE**   UNDER CONSTRUCTION
BUILDER   C.E.M.A. (Center for Advanced Marine Studies)
OWNER/OPERATOR   Institute Français Petroleum/CNEXO.

# Appendices

| | | | | |
|---|---|---|---|---|
| MAX. OPERATING DEPTH | 1,970 ft. | PILOTS 6 | | DIVERS 4 |
| COLLAPSE DEPTH | 3,800 ft. | VIEWPORTS n. a. | | |
| HULL SIZE | 144 in. | MANIPULATORS | | none |
| LENGTH 82 ft. BEAM | 22 ft. | PROPULSION | | 2 stern props |
| HEIGHT 26 ft. WEIGHT 500,000 lbs. | | POWER diesel/lead acid 1,200 kwh. | | |
| MATERIAL | steel | CRUISE SPEED | | 3.6 knots |
| PAYLOAD | n. a. | MAX. SPEED (sfc.) | | 7 knots |
| LIFE SUPPORT | up to 10 days max. | | | |

MAJOR EQUIPMENT   Lock-out capability to 2,000 ft.; cruise range surface, 400 miles; extended endurance—10 days with fuel cell.

ESTIMATED COST   $6.5 million

STATUS   Construction postponed. Due for launch in 1973.

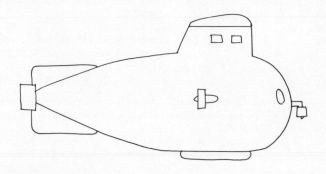

## ASHERAH   LAUNCHED 1964

BUILDER   General Dynamics/Electric Boat Div., Groton, Conn.

OWNER/OPERATOR   Technoceans Inc., New York, N.Y.

| | | | |
|---|---|---|---|
| MAX. OPERATING DEPTH | 600 ft. | PILOTS 1 | OBSERVERS 1 |
| MAX. DEPTH ACHIEVED | 600 ft. | VIEWPORTS 6 forward/down, port, | |
| HULL SIZE | 60 in. | stbd. | |
| LENGTH 17 ft. BEAM | 7.6 ft. | MANIPULATORS | none |
| HEIGHT 7.6 ft. WEIGHT | 8,500 lbs. | PROPULSION (2) 2-hp DC motors, | |
| MATERIAL | steel | side mounted | |
| PAYLOAD | 350 lbs. | POWER lead acid battery 24 kwh. | |
| LIFE SUPPORT | 48 hrs. max. | CRUISE SPEED 1 knot for 8 hrs. | |
| | | MAX. SPEED 3 knots for 1.5 hrs. | |

MAJOR EQUIPMENT   Magnesyn compass, echo sounder, UQC, depth gauge, CB radio, external lighting

ESTIMATED COST   $25,000

203

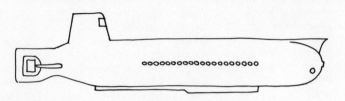

### AUGUSTE PICCARD (PX-8)   LAUNCHED 1963
BUILDER  Jacques Piccard/Giavanola, Switzerland
OWNER/OPERATOR  Chicago Bridge and Iron Corp., Oakbrook, Ill.

| | | | |
|---|---|---|---|
| NUMBER OF DIVES | 1,100 | PAYLOAD | 20,000 lbs. |
| TOTAL DIVE HOURS | 12,000 | LIFE SUPPORT | 2,160 hrs. max. |
| MAX. OPERATING DEPTH | 2,500 ft. | PILOTS 5 | PASSENGERS 40 |
| MAX. DEPTH ACHIEVED | 1,060 ft. | VIEWPORTS 46 | |
| COLLAPSE DEPTH | 4,500 ft. | MANIPULATORS | none |
| HULL SIZE | 111 in. | PROPULSION | (1) 80-hp DC motor |
| LENGTH 93 ft. BEAM | 19 ft. | POWER lead acid battery 600 kwh. | |
| HEIGHT 24 ft. WEIGHT | 165 tons | CRUISE SPEED | 6 knots for 10 hrs. |
| MATERIAL | steel | MAX. SPEED | 6.3 knots for 7 hrs. |

MAJOR EQUIPMENT  Artificial horizon, inclinometer, TV with monitors for
          passengers, surface radio
ESTIMATED COST  $2.5 million
STATUS  In storage in Marseilles, France, after use at Swiss Fair, 1964.
     Awaiting disposition by Chicago Bridge and Iron Corp.

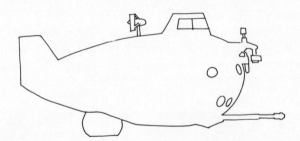

### BEAVER IV (Rough Neck)   LAUNCHED September 1968
BUILDER  North American Rockwell Corp., Seal Beach, Calif.
OWNER/OPERATOR  Same

| | | | |
|---|---|---|---|
| NUMBER OF DIVES | 72 | COLLAPSE DEPTH | 4,000 ft. |
| TOTAL DIVE HOURS | 255 | HULL SIZE 85 in. forward sphere | |
| MAX. OPERATING DEPTH | 2,000 ft. | 60 in. aft sphere | |
| MAX. DEPTH ACHIEVED | 2,000 ft. | LENGTH 24 ft. BEAM | 9.5 ft. |

HEIGHT 8.5 ft.  WEIGHT 28,000 lbs.
MATERIAL           HY-100 steel
PAYLOAD           2,000 lbs.
LIFE SUPPORT     180 hrs. max.
PILOTS 2      OBSERVERS 3
VIEWPORTS  9 forward, 2 in hatch

MANIPULATORS  (2) 250 lb.; 6-ft.
               reach with tools
PROPULSION   (3) 5-hp DC motors
POWER   lead acid battery 44 kwh.
CRUISE SPEED  2.5 knots for 8 hrs.
MAX. SPEED    5 knots for 0.3 hrs.

MAJOR EQUIPMENT  Gyrocompass, altitude/depth sonar, TV on pan and tilt, 70mm still camera, CTFM sonar, UQC, surface (VHF) radio, diver lockout
ESTIMATED COST  $4 million
STATUS  In storage.

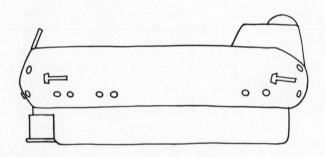

## BEN FRANKLIN (PX-15)   LAUNCHED 1968
BUILDER  Jacques Piccard/Giavanola, Switzerland
OWNER/OPERATOR  Grumman Aerospace Corp., Bethpage, L.I., N.Y.

NUMBER OF DIVES         70
TOTAL DIVE HOURS     1,150
MAX. OPERATING DEPTH  2,000 ft.
MAX. DEPTH ACHIEVED   2,000 ft.
COLLAPSE DEPTH      4,000 ft.
HULL SIZE           120 in.
LENGTH 48 ft.  BEAM    21 ft.
HEIGHT 20 ft.  WEIGHT 286,000 lbs.
MATERIAL              steel
PAYLOAD          9,345 lbs.

LIFE SUPPORT     6,048 hrs. max.
PILOTS 2      OBSERVERS 4
VIEWPORTS 27
MANIPULATORS        none
PROPULSION  (4) 25-hp, var. frequency
POWER  lead acid battery 756 kwh.
CRUISE SPEED       2.5 knots
MAX. SPEED        4 knots

MAJOR EQUIPMENT  Gyrocompass, UQC, sonar tracking equipment, TV, still camera, water sampler, hydrophones, exterior lighting, surface radio
ESTIMATED COST  $3 million
STATUS  Damaged on reef in Bahamas, 1970; completely overhauled in 1971 and components stored.

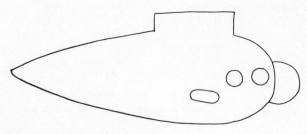

**BENTHOS V**     LAUNCHED   1963
BUILDER   Lear Sigler Corp.
OWNER/OPERATOR   Same

| | | | |
|---|---|---|---|
| NUMBER OF DIVES | 5 | LIFE SUPPORT | 6 hrs. max. |
| MAX. OPERATING DEPTH | 600 ft. | PILOTS 1 | OBSERVERS 1 |
| MAX. DEPTH ACHIEVED | n. a. | VIEWPORTS 2 | |
| COLLAPSE DEPTH | n. a. | MANIPULATORS | none |
| HULL SIZE | 44 in. | PROPULSION | (2) 1-hp. DC motors |
| LENGTH 11 ft. BEAM | 4 ft. | POWER | silver cadmium battery |
| HEIGHT 5 ft. WEIGHT | 4,000 lbs. | CRUISE SPEED | 2 knots for 2 hrs. |
| MATERIAL | steel | MAX. SPEED | 3 knots for 1 hr. |
| PAYLOAD | 400 lbs. | | |

MAJOR EQUIPMENT   Doppler navigation
ESTIMATED COST   n. a.
STATUS   In storage; never operational.

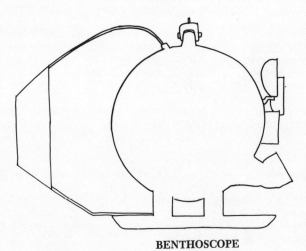

**BENTHOSCOPE**

**BENTHOSCOPE**    LAUNCHED    1948
BUILDER    Watson Stillman Co., N.J.
OWNER/OPERATOR    Otis Barton

| | | | |
|---|---|---|---|
| NUMBER OF DIVES | 10 | LIFE SUPPORT | 24 hrs. max. |
| TOTAL DIVE HOURS | n. a. | PILOTS 1 | |
| MAX. OPERATING DEPTH | 10,000 ft. | VIEWPORTS | 2—fused quartz |
| MAX. DEPTH ACHIEVED | 4,480 ft. | MANIPULATORS | none |
| COLLAPSE DEPTH | 15,000 ft. | PROPULSION | none—on cable to |
| HULL SIZE | 54 in. | | surface |
| LENGTH 6 ft. BEAM | 5 ft. | POWER | generator from surface |
| HEIGHT 6 ft. WEIGHT | 7,000 lbs. | CRUISE SPEED | 2 knots on tether |
| MATERIAL | steel | | cable |
| PAYLOAD | none | | |

MAJOR EQUIPMENT    1,500-watt spotlight, hardwire phone, camera, wooden wheels for bottom rolling
ESTIMATED COST    $16,000
STATUS    Never used after dive to 4,480 ft. in 1949.

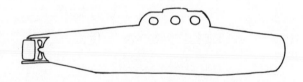

**CUBMARINE (PC 3A-1)**    LAUNCHED    1962
BUILDER    Perry Submarine Builders, Riviera Beach, Fla.
OWNER/OPERATOR    U.S. Army/Kentron, Hawaii Ltd.

| | | | |
|---|---|---|---|
| NUMBER OF DIVES | 600 | PAYLOAD | 750 lbs.; extra 340 lbs. |
| TOTAL DIVE HOURS | 600 | LIFE SUPPORT | 20 hrs. max. |
| MAX. OPERATING DEPTH | 300 ft. | PILOTS 1 | OBSERVERS 1 |
| MAX. DEPTH ACHIEVED | 300 ft. | VIEWPORTS | 13 in conning tower |
| COLLAPSE DEPTH | 500 ft. | MANIPULATORS | none |
| HULL SIZE | 56 in. | PROPULSION | (1) 5-hp DC motor |
| LENGTH 18.5 ft. BEAM | 3.5 ft. | POWER | lead acid battery 7.5 kwh. |
| HEIGHT 5 ft. WEIGHT | 4,790 lbs. | CRUISE SPEED | 1.5 knots for 8 hrs. |
| MATERIAL | steel | MAX. SPEED | 4 knots for 1 hr. |

MAJOR EQUIPMENT    Magnetic compass, underwater telephone
ESTIMATED COST    $30,000
STATUS    Diving in Marshall Islands to recover missile components for U.S. Army.

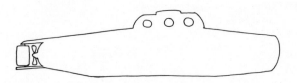

**CUBMARINE (PC 3A-2)**     LAUNCHED   1965
BUILDER   Perry Submarine Builders, Riviera Beach, Fla.
OWNER/OPERATOR   U.S. Air Force Pacific/Kentron, Hawaii Ltd.

| | | | |
|---|---|---|---|
| NUMBER OF DIVES | 350 | LIFE SUPPORT | 20 hrs. max. |
| TOTAL DIVE HOURS | 300 | PILOTS 1 | OBSERVERS 1 |
| MAX. OPERATING DEPTH | 300 ft. | VIEWPORTS | 13 in conning tower |
| MAX. DEPTH ACHIEVED | 300 ft. | MANIPULATORS | none |
| COLLAPSE DEPTH | 500 ft. | PROPULSION | (1) 5-hp DC motor; |
| HULL SIZE | 56 in. | | 1 bow thruster |
| LENGTH 18.5 ft. BEAM | 3.5 ft. | POWER | lead acid battery 7.5 kwh. |
| HEIGHT 5 ft. WEIGHT | 4,790 lbs. | CRUISE SPEED | 1.5 knots for 8 hrs. |
| MATERIAL | steel | MAX. SPEED | 4 knots for 1 hr. |
| PAYLOAD 750 lbs.; extra 340 lbs. | | | |

MAJOR EQUIPMENT   Magnetic compass, underwater telephone
ESTIMATED COST   $30,000
STATUS   Diving in Marshall Islands for U.S. Air Force to recover missile
        parts.

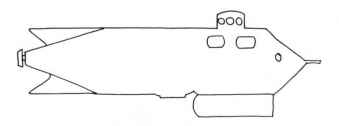

**DEEP DIVER (PLC-4)**     LAUNCHED   1968
BUILDER   Perry Submarine Builders, Riviera Beach, Fla.
OWNER/OPERATOR   Perry Oceanographics Inc.

| | | | | |
|---|---|---|---|---|
| NUMBER OF DIVES | 150 | COLLAPSE DEPTH | | 2,000 ft. |
| TOTAL DIVE HOURS | 300 | HULL SIZE | | 54 in. |
| MAX. OPERATING DEPTH | 1,335 ft. | LENGTH 23 ft. | BEAM | 5.5 ft. |
| MAX. DEPTH ACHIEVED | 1,505 ft. | HEIGHT 6 ft. | WEIGHT | 16,500 lbs. |

| | | | |
|---|---|---|---|
| MATERIAL | steel | PROPULSION | (1) 10-hp DC stern motor |
| PAYLOAD | 2,000 lbs. | | |
| LIFE SUPPORT | 50 hrs. max. | POWER | lead acid battery 23 kwh. |
| PILOTS 1 | OBSERVERS 3 | CRUISE SPEED | 1.5 knots for 10 hrs. |
| VIEWPORTS | (17) 8-in. diameter | MAX. SPEED | 3.5 knots for 5 hrs. |
| MANIPULATORS | rock drill | | |

MAJOR EQUIPMENT   Magnesyn compass, echo sounder, underwater telephone, CB radio, lock-out chamber, exploration package (optional), cameras, manipulator rock drill

ESTIMATED COST   n. a.

STATUS   Stationed in Morgan City, La., for charter or sale to oil companies.

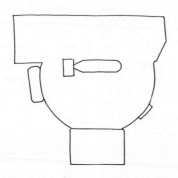

## DEEP JEEP   LAUNCHED   1964

BUILDER   Naval Ordnance Test Station, China Lake, Calif.

OWNER/OPERATOR   Scripps Institution of Oceanography, La Jolla, Calif.

| | | | |
|---|---|---|---|
| NUMBER OF DIVES | 100 | LIFE SUPPORT | 100 hrs. max. |
| TOTAL DIVE HOURS | 250 | PILOTS 1 | OBSERVERS 1 |
| MAX. OPERATING DEPTH | 2,000 ft. | VIEWPORTS | viewing optics |
| MAX. DEPTH ACHIEVED | 2,000 ft. | MANIPULATORS | none |
| COLLAPSE DEPTH | n. a. | PROPULSION | (2) ¾-hp tiltable motors |
| HULL SIZE | 68 in. | | |
| LENGTH 10 ft. BEAM | 8.5 ft. | POWER | lead acid battery 7 kwh. |
| HEIGHT 8 ft. WEIGHT | 9,600 lbs. | CRUISE SPEED | 1 knot for 6 hrs. |
| MATERIAL | steel | MAX. SPEED | 2 knots for 2 hrs. |
| PAYLOAD | n. a. | | |

MAJOR EQUIPMENT   UQC, external lights, fixed optical tube for pilot/ob-
server viewing

ESTIMATED COST   n. a.

STATUS   Donated to Scripps Institution of Oceanography in 1966. In
storage.

# Appendices

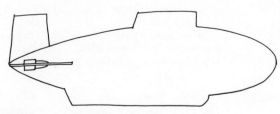

**DEEP QUEST**     LAUNCHED   1967
BUILDER    Lockheed Missiles and Space Corp., Sunnyvale, Calif.
OWNER/OPERATOR    Lockheed, San Diego, Calif.

| | | | |
|---|---|---|---|
| NUMBER OF DIVES | 150 | LIFE SUPPORT | 192 hrs. max. |
| TOTAL DIVE HOURS | 625 | PILOTS 2 | OBSERVERS 2 |
| MAX. OPERATING DEPTH | 8,000 ft. | VIEWPORTS | 2 forward/down |
| MAX. DEPTH ACHIEVED | 8,350 ft. | MANIPULATORS | 1—general purpose |
| COLLAPSE DEPTH | 13,000 ft. | PROPULSION | (2) 7.5-hp stern; |
| HULL SIZE | (2) 84-in. spheres | | (2) 7.5-hp thrusters |
| LENGTH 40 ft. BEAM | 19 ft. | POWER | lead acid battery 24.5 kwh. |
| HEIGHT 13 ft. WEIGHT 110,000 lbs. | | CRUISE SPEED | 2 knots for 24 hrs. |
| MATERIAL | maraging steel | MAX. SPEED | 4.5 knots for 10 hrs. |
| PAYLOAD | 5,000 lbs. | | |

MAJOR EQUIPMENT   Gyrocompass, TV system, CTFM sonar, altitude/depth
          sonar, ship control computer, cameras and lights, corer
ESTIMATED COST   n. a.
STATUS   Diving for Lehigh University off San Diego, and on Lockheed
         dives.

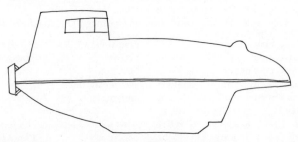

**DEEPSTAR 2000**     LAUNCHED   December 1969
BUILDER    Westinghouse Electric Corp.
OWNER/OPERATOR    Westinghouse Research and Development/Ocean Re-
         search Lab, Annapolis, Md.

| | | | |
|---|---|---|---|
| NUMBER OF DIVES | 60 | MAX. DEPTH ACHIEVED | 2,000 ft. |
| TOTAL DIVE HOURS | 150 | COLLAPSE DEPTH | 4,000 ft. |
| MAX. OPERATING DEPTH | 2,000 ft. | HULL SIZE | 60 in. |

LENGTH 20 ft.   BEAM   7.5 ft.
HEIGHT 8 ft.   WEIGHT   17,500 lbs.
MATERIAL   HY-80 steel
PAYLOAD   1,500 lbs.
LIFE SUPPORT   144 hrs. max.
PILOTS 1   OBSERVERS 2
VIEWPORTS   2 forward, 1 camera port

MANIPULATORS   1 hydraulic—general purpose
PROPULSION   (1) 10-hp DC drive; 2 hydraulic stern; 4 thrusters (shrouded)
POWER   lead acid battery 26.5 kwh.
CRUISE SPEED   1 knot for 8 hrs.
MAX. SPEED   3 knots for 4 hrs.

MAJOR EQUIPMENT   Gyrocompass, altitude/depth sonar, still and movie cameras, lights, (2) 250-watt lights on 10-ft. booms, sediment sampler, temperature and water velocity
ESTIMATED COST   $1 million
STATUS   In storage in Annapolis, Md.

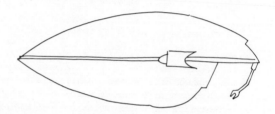

**DEEPSTAR 4000**   LAUNCHED   May 1965
BUILDER   Office Français de Recherches Sous-Marine and Westinghouse Electric Corp.
OWNER/OPERATOR   Westinghouse Ocean Research and Engineering Center, Annapolis, Md.

NUMBER OF DIVES   544
TOTAL DIVE HOURS   2,000 +
MAX. OPERATING DEPTH   4,000 ft.
MAX. DEPTH ACHIEVED   4,135 ft.
COLLAPSE DEPTH   7,600 ft.
HULL SIZE   78 in.
LENGTH 18 ft.   BEAM   11.5 ft.
HEIGHT 7 ft.   WEIGHT   18,300 lbs.
MATERIAL   HY-80 steel
PAYLOAD   600 lbs.
LIFE SUPPORT   144 hrs. max.

PILOTS 1   OBSERVERS 2
VIEWPORTS   2 forward, 1 camera port
MANIPULATORS   1 hydraulic—sampling
PROPULSION   (2) 5-hp AC motors on side
POWER   lead acid battery 48 kwh.
CRUISE SPEED   1.5 knots for 6 hrs.
MAX. SPEED   3 knots for 4 hrs.

MAJOR EQUIPMENT   Gyrocompass, echo sounder, tracking pinger, still and movie cameras, lights, UQC, FM surface radio, current meter, temperature gauge, instrument brow
ESTIMATED COST   $1 million
STATUS   In storage in North Carolina.

# Appendices

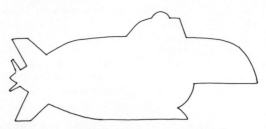

**DEEPSTAR 20,000**    CONSTRUCTION HALTED    1970
BUILDER   Westinghouse Ocean Research and Engineering Center, Annapolis, Md.
OWNER/OPERATOR   Same

| | | | |
|---|---|---|---|
| MAX. OPERATING DEPTH | 20,000 ft. | VIEWPORTS | 2 forward, 1 camera port |
| COLLAPSE DEPTH | 30,000 ft. | | |
| HULL SIZE | 80 in. | MANIPULATORS | 1 hydraulic—general purpose |
| LENGTH 36 ft.   BEAM | 10.2 ft. | | |
| HEIGHT 13.5 ft.   WEIGHT | 84,000 lbs. | PROPULSION | (1) 10-hp AC motor; Varivec propeller |
| MATERIAL | HY-140 steel | | |
| PAYLOAD | 2,000 lbs. | POWER | silver zinc battery 160 kwh. |
| LIFE SUPPORT | 216 hrs. max. | CRUISE SPEED | 2 knots for 8 hrs. |
| PILOTS 1 | OBSERVERS 2 | MAX. SPEED | 3 knots for 4 hrs. |

MAJOR EQUIPMENT   Gyrocompass, forward scan sonar, UQC, still and movie cameras, FM surface radio, altitude/depth sonar
ESTIMATED COST   $5 million
STATUS   Component construction 80% complete but halted in 1970. Now in storage.

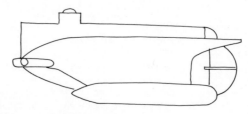

**DEEPVIEW**    IN TRIALS    1971
BUILDER   U.S. Naval Underseas R & D Center
OWNER/OPERATOR   Same

| | | | | |
|---|---|---|---|---|
| NUMBER OF DIVES | trials | HULL SIZE | | 44 in. |
| MAX. OPERATING DEPTH | 1,500 ft. | LENGTH 16.5 ft. | BEAM | 6 ft. |
| MAX. DEPTH ACHIEVED | 10 ft. | HEIGHT 6.5 ft. | WEIGHT | 12,000 lbs. |
| COLLAPSE DEPTH | 6,500 ft. | MATERIAL | | glass/HY-100 steel |

| | | | |
|---|---|---|---|
| PAYLOAD | 500 lbs. | PROPULSION | (2) 12-hp stern; 1 lat- |
| LIFE SUPPORT | 24 hrs. max. | | eral, 1 vertical |
| PILOTS 1 | OBSERVERS 1 | POWER | lead acid battery 20 kwh. |
| VIEWPORTS | glass hemi-head, 44-in. | CRUISE SPEED | 1.5 knots for 6 hrs. |
| | diameter | MAX. SPEED | 5 knots for 1 hr. |
| MANIPULATORS | none | | |

MAJOR EQUIPMENT  Underwater telephone (UQC at 9 KHz), 50 KHz pinger, inside cameras optional

ESTIMATED COST  n. a.

STATUS  Shallow water trials conducted during 1971; certification applied for.

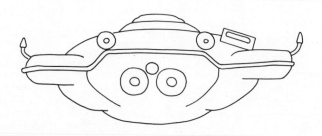

## DIVING SAUCER (SP 300)
### SOUCOUPE ("DENISE")  LAUNCHED  1959
BUILDER  Office Français de Recherches Sous-Marine, Marseilles, France
OWNER/OPERATOR  OFRS/Cousteau

| | | | |
|---|---|---|---|
| NUMBER OF DIVES | 750 | PILOTS 1 | OBSERVERS 1 |
| TOTAL DIVE HOURS | 2,000 | VIEWPORTS | 2 forward, 1 camera |
| MAX. OPERATING DEPTH | 1,350 ft. | | port |
| MAX. DEPTH ACHIEVED | 1,350 ft. | MANIPULATORS | 1 hydraulic—gen- |
| COLLAPSE DEPTH | 3,300 ft. | | eral purpose |
| HULL SIZE | 65 in. by 49 in. | PROPULSION | (1) 2-hp DC motor |
| LENGTH 9 ft. BEAM | 9 ft. | | driving 2 water jets |
| HEIGHT 5 ft. WEIGHT | 8,400 lbs. | POWER | lead acid battery 13 kwh. |
| MATERIAL | steel | CRUISE SPEED | 0.6 knots for 4 hrs. |
| PAYLOAD | 300 lbs. | MAX. SPEED | 1 knot for 2 hrs. |
| LIFE SUPPORT | 48 hrs. max. | | |

MAJOR EQUIPMENT  Gyro, echo sounder, underwater telephone, still and movie cameras, movie light on 10-ft. boom

ESTIMATED COST  $150,000

STATUS  Diving occasionally for OFRS/Cousteau.

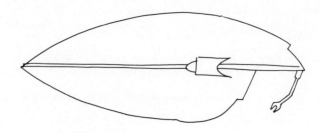

**DIVING SAUCER (SP 3000)**     LAUNCHED  July 1970
BUILDER   Office Français de Recherches Sous-Marine, Marseilles, France
OWNER/OPERATOR   OFRS/Cousteau

| | | | |
|---|---|---|---|
| MAX. OPERATING DEPTH | 9,500 ft. | PILOTS  1 | OBSERVERS  2 |
| COLLAPSE DEPTH | 15,000 ft. | VIEWPORTS  2 forward, 1 camera | |
| HULL SIZE | 78 in. | port | |
| LENGTH  20 ft.  BEAM | 10 ft. | MANIPULATORS  1—general purpose | |
| HEIGHT  7 ft.  WEIGHT | 24,000 lbs. | PROPULSION    (2) 5-hp AC motors | |
| MATERIAL | vasco 90 | POWER    lead acid battery 25 kwh. | |
| PAYLOAD | 500 lbs. | CRUISE SPEED    2 knots for 10 hrs. | |
| LIFE SUPPORT | 144 hrs. max. | MAX. SPEED    3 knots for  5 hrs. | |

MAJOR EQUIPMENT   Gyro, echo sounder, underwater radio, still and movie
                 cameras, external lights
ESTIMATED COST   n. a.
STATUS   Completed sea trials late 1970.

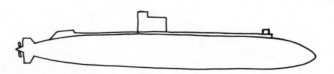

**DOLPHIN (AGSS-555)**     LAUNCHED   1968
BUILDER   U.S. Navy, Portsmouth Naval Shipyard, Portsmouth, N.H.
OWNER/OPERATOR   U.S. Navy COMSUBDEVGRU ONE, San Diego, Calif.

| | | | |
|---|---|---|---|
| NUMBER OF DIVES | 35 | MAX. DEPTH ACHIEVED | classified |
| TOTAL DIVE HOURS | 250 | COLLAPSE DEPTH | classified |
| MAX. OPERATING DEPTH | classified | HULL SIZE | 19 ft. |

LENGTH 165 ft.    BEAM        19 ft.    MANIPULATORS                none
HEIGHT  20 ft.    WEIGHT    900 tons    PROPULSION   diesel/electric—
MATERIAL              HY-80 steel                      1,500 hp.
PAYLOAD                 12 tons    POWER                silver zinc/diesel
LIFE SUPPORT    24 hrs. max.    CRUISE SPEED    5 knots for 16 hrs.
CREW  22         OBSERVERS  5    MAX. SPEED      15 knots for  5 hrs.
VIEWPORTS                 n. a.

MAJOR EQUIPMENT   Primarily sonar equipment for test platform; advanced
design of pressure-compensated sonars to save weight.

ESTIMATED COST   n. a.

STATUS   Completed cruise to Kodiak, Alaska, in fall of 1971. Scheduled for
continuing operations out of San Diego, 1972.

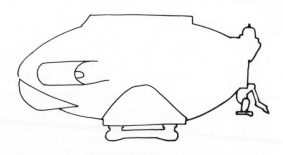

## DOWB (DEEP OCEAN WORK BOAT)        LAUNCHED   1968

BUILDER   AC Division, General Motors, Santa Barbara, Calif.

OWNER/OPERATOR   Same

NUMBER OF DIVES              115    PILOTS  1              OBSERVERS  1
TOTAL DIVE HOURS            300    VIEWPORTS   special   optical   view
MAX. OPERATING DEPTH   6,500 ft.            system
MAX. DEPTH ACHIEVED    6,500 ft.    MANIPULATORS   1  hydraulic—gen-
COLLAPSE DEPTH           10,000 ft.            eral purpose
HULL SIZE                     82 in.    PROPULSION  (2)  2-hp  horizontal;
LENGTH  18 ft.   BEAM        8.8 ft.                (2)  2-hp  vertical
HEIGHT 10.5 ft.   WEIGHT  20,000 lbs.    POWER     lead acid battery 40 kwh.
MATERIAL              HY-100 steel    CRUISE SPEED        1 knot for 26 hrs.
PAYLOAD                  600 lbs.    MAX. SPEED        2 knots for  6 hrs.
LIFE SUPPORT     180 hrs. max.

MAJOR EQUIPMENT   Gyrocompass, CTFM sonar, TV, still and movie cameras, UQC, surface radio

ESTIMATED COST   $2 million

STATUS   In storage.

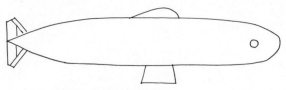

**DSRV-1**  LAUNCHED  January 1970
BUILDER  Lockheed Missiles and Space Corp., Sunnyvale, Calif.
OWNER/OPERATOR  U.S. Navy Ship Systems Com/COMSUBDEVGRU
ONE, San Diego, Calif.

| | | | |
|---|---|---|---|
| NUMBER OF DIVES | 30 | LIFE SUPPORT | 310 hrs. max. |
| TOTAL DIVE HOURS | 100 | PILOTS 3 | RESCUEES 24 |
| MAX. OPERATING DEPTH | 5,000 ft. | VIEWPORTS 3 | |
| MAX. DEPTH ACHIEVED | 3,500 ft. | MANIPULATORS | 1 electrohydraulic |
| COLLAPSE DEPTH | 7,500 ft. | PROPULSION | (1) 15-hp main; |
| HULL SIZE | 3 spheres, 90 in. | | (4) 75-hp ducted |
| LENGTH 50 ft. BEAM | 8 ft. | | thrusters |
| HEIGHT 12 ft. WEIGHT | 73,900 lbs. | POWER | silver zinc battery 58 kwh. |
| MATERIAL | HY-140 steel | CRUISE SPEED | 3 knots for 12 hrs. |
| PAYLOAD | 4,320 lbs. | MAX. SPEED | 4.5 knots for 3 hrs. |

MAJOR EQUIPMENT  Integrated control and display (ICAD), 3 TV cameras,
4 monitors, CTFM sonar, altitude/depth sonar, Dop-
pler navigator, control computer, UQC/VHF radio,
still camera, movie camera, lights, transponder navigator
ESTIMATED COST  $50 million
STATUS  Underwent U.S. Navy sea trials during 1971. *DSRV-2* launched
mid-1971, identical to *DSRV-1*. Operating in trials with mother
ship, U.S.S. BN666 *Hawkbill*.

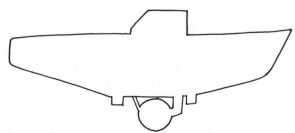

**FNRS-3**  LAUNCHED  1953
BUILDER  French Navy
OWNER/OPERATOR  French Navy, Toulon, France

| | | | |
|---|---|---|---|
| NUMBER OF DIVES | 40 | LIFE SUPPORT | 100 hrs. max. |
| TOTAL DIVE HOURS | n. a. | PILOTS 1 | OBSERVERS 1 |
| MAX. OPERATING DEPTH | 13,000 ft. | VIEWPORTS 2 | |
| MAX. DEPTH ACHIEVED | 13,000 ft. | MANIPULATORS | none |
| COLLAPSE DEPTH | 20,000 ft. | PROPULSION | (2) 1-hp motors |
| HULL SIZE | 78 in. | POWER | lead acid battery |
| LENGTH 52 ft. BEAM | 14 ft. | CRUISE SPEED | 0.5 knots for 8 hrs. |
| HEIGHT 26 ft. WEIGHT | 56,000 lbs. | MAX. SPEED | 1 knot for 2 hrs. |
| MATERIAL | steel | | |

MAJOR EQUIPMENT Lights, speedometer, pressure gauge
ESTIMATED COST n. a.
STATUS Has not been diving since 1960.

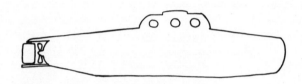

**GASPERGOU (PC-3X)**     LAUNCHED 1960
BUILDER Perry Submarine Builders, Riviera Beach, Fla.
OWNER/OPERATOR Applied Research Labs., University of Texas, Austin, Tex.

| | | | |
|---|---|---|---|
| NUMBER OF DIVES | 500 (150 with Texas) | MATERIAL | steel |
| | | PAYLOAD | 100 lbs. |
| TOTAL DIVE HOURS | 500 (as of January 1970) | LIFE SUPPORT | 48 hrs. max. |
| | | PILOTS 1 | OBSERVERS 1 |
| MAX. OPERATING DEPTH | 150 ft. | VIEWPORTS | 13 in conning tower |
| MAX. DEPTH ACHIEVED | (in test) 225 ft. | MANIPULATORS | none |
| | | PROPULSION | (1) 4-hp DC motor, active rudder |
| COLLAPSE DEPTH | 500 ft. | | |
| HULL SIZE | 56 in. | POWER | lead acid battery 11 kwh. |
| LENGTH 18.5 ft. BEAM | 3.5 ft. | CRUISE SPEED | 2 knots for 6 hrs. |
| HEIGHT 5 ft. WEIGHT | 4,700 lbs. | MAX. SPEED | 4 knots for 3 hrs. |

MAJOR EQUIPMENT Magnesyn compass, echo sounder, underwater telephone, forward looking sonar, CB radio, acoustic transponder
ESTIMATED COST $30,000
STATUS Diving occasionally in Lake Travis, Tex.

# Appendices

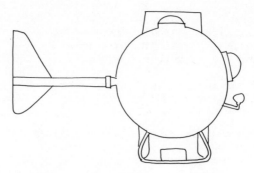

**GUPPY**    LAUNCHED   April 1970
BUILDER   Sun Shipbuilding and Dry Dock Corp., Chester, Pa.
OWNER/OPERATOR   Same

| | | | |
|---|---|---|---|
| NUMBER OF DIVES | 50 | LIFE SUPPORT | 56 hrs. max. |
| TOTAL DIVE HOURS | 85 | PILOTS 1 | OBSERVERS 1 |
| MAX. OPERATING DEPTH | 1,000 ft. | VIEWPORTS | (1) 16-in. in hatch; |
| MAX. DEPTH ACHIEVED | 1,100 ft. | | (2) 8-in. forward |
| COLLAPSE DEPTH | 2,700 ft. | MANIPULATORS | none |
| HULL SIZE | 66 in. | PROPULSION | (2) 10-hp, 440 vac |
| LENGTH 11 ft. BEAM | 8 ft. | POWER | electric cable 35 kwh. |
| HEIGHT 7.5 ft. WEIGHT | 5,800 lbs. | CRUISE SPEED | 1 knot |
| MATERIAL | HY-100 steel | MAX. SPEED | 3 knots |
| PAYLOAD | n. a. | | |

MAJOR EQUIPMENT   Vehicle on 1,200-ft. tether, hardwire phone, gyro, corer, current meters, (4) 1,000-watt external lights
ESTIMATED COST   $95,000
STATUS   Completed dives in Gulf of Alaska, 1970; dived in Santa Barbara Channel, 1971.

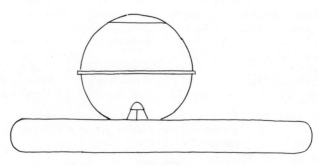

**HIKINO**

## HIKINO (THS-1)    UNDER CONSTRUCTION
BUILDER   U.S. Navy/NUC, Hawaii
OWNER/OPERATOR   Same

| | | | |
|---|---|---|---|
| MAX. OPERATING DEPTH | 600 ft. | MATERIAL | acrylic |
| MAX. DEPTH ACHIEVED | 600 ft. | LIFE SUPPORT | 36 hrs. max. |
| COLLAPSE DEPTH | n. a. | PILOTS 1 | OBSERVERS 1 |
| HULL SIZE | 66 in. | VIEWPORTS | acrylic sphere |
| LENGTH | 18.7 ft. | | |

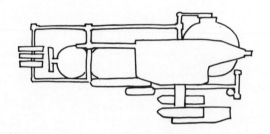

## JOHNSON-SEA-LINK    LAUNCHED   January 1971
BUILDER   Edwin A. Link
OWNER/OPERATOR   Smithsonian Institution, Washington, D.C.

| | | | |
|---|---|---|---|
| NUMBER OF DIVES | sea trials | LIFE SUPPORT | 60 hrs. max. |
| MAX. OPERATING DEPTH | 2,000 ft. | PILOTS 1 | OBSERVERS 4 |
| MAX. DEPTH ACHIEVED | trials | VIEWPORTS | acrylic clear sphere |
| COLLAPSE DEPTH | 5,000 ft. | MANIPULATORS | none |
| HULL SIZE | 66 in. | PROPULSION | 6 DC motor thrusters |
| LENGTH 23 ft.  BEAM | 7.5 ft. | POWER | lead acid battery 30 kwh. |
| HEIGHT 9 ft.  WEIGHT | 19,000 lbs. | CRUISE SPEED | 2 knots for 48 hrs. |
| MATERIAL | acrylic and steel | MAX. SPEED | 4 knots for 24 hrs. |
| PAYLOAD | 500 lbs. | | |

MAJOR EQUIPMENT   Lock-out chamber in 48 in. by 8 ft. cylinder, Doppler
navigator, sonar

ESTIMATED COST   n. a.

STATUS   Undergoing sea trials. To be used for scientific diving for the
Smithsonian Institution.

# Appendices

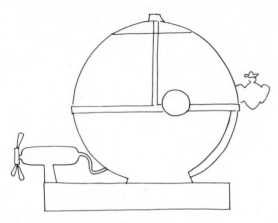

**KUMUKAHI**   LAUNCHED  1969
BUILDER   Oceanic Foundation, Honolulu, Hawaii
OWNER/OPERATOR   Same

| | | | |
|---|---|---|---|
| NUMBER OF DIVES | experimental | MATERIAL | acrylic |
| MAX. OPERATING DEPTH | 300 ft. | LIFE SUPPORT | 8 hrs. max. |
| LENGTH | 6 ft. | PILOTS 1 | OBSERVERS 1 |
| WEIGHT | 3,700 lbs. | VIEWPORTS | acrylic clear sphere |

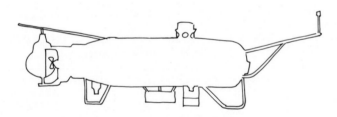

**KUROSHIO II**   LAUNCHED  1960
BUILDER   Japanese Steel and Tube Co., Hokkaido, Japan
OWNER/OPERATOR   University of Hokkaido

| | | | |
|---|---|---|---|
| NUMBER OF DIVES | 135 | LENGTH 36 ft. | BEAM 7 ft. |
| TOTAL DIVE HOURS | n. a. | HEIGHT 10 ft. | WEIGHT 23,000 lbs. |
| MAX. OPERATING DEPTH | 650 ft. | MATERIAL | steel |
| MAX. DEPTH ACHIEVED | 650 ft. | PAYLOAD | n. a. |
| COLLAPSE DEPTH | 1,200 ft. | LIFE SUPPORT | 96 hrs. max. |
| HULL SIZE | 50 in. | PILOTS 2 | OBSERVERS 2 |

VIEWPORTS 16

MANIPULATORS none

PROPULSION (1) 4-hp AC motor

POWER 1,900-ft. cable from surface

CRUISE SPEED 1 knot

MAX. SPEED 2 knots

MAJOR EQUIPMENT Hardwire phone, forward and down sonar, hydrophone, current meter

ESTIMATED COST n. a.

STATUS Inactive.

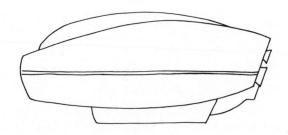

## MINI SUBS (SEA FLEAS I AND II)

LAUNCHED 1969

BUILDER Sud Aviation, France

OWNER/OPERATOR OFRS/Cousteau

NUMBER OF DIVES 200 each

TOTAL DIVE HOURS 300

MAX. OPERATING DEPTH 2,000 ft.

MAX. DEPTH ACHIEVED 2,000 ft.

COLLAPSE DEPTH 4,000 ft.

HULL SIZE 39 in.

LENGTH 9.5 ft. BEAM 6.3 ft.

HEIGHT 4.4 ft. WEIGHT 5,300 lbs.

MATERIAL steel

PAYLOAD 100 lbs.

LIFE SUPPORT 24 hrs. max.

PILOTS 1

VIEWPORTS 1 forward, 1 camera port

MANIPULATORS 1 hydraulic—general purpose

PROPULSION electric motor/pump/water jet

POWER lead acid battery 6.8 kwh.

CRUISE SPEED 0.8 knots for 2 hrs.

MAX. SPEED 1.1 knots for 1.5 hrs.

MAJOR EQUIPMENT Gyro, echo sounder, pinger, 2 movie cameras, radio, underwater telephone

ESTIMATED COST n. a.

STATUS Both in use by OFRS/Cousteau for exploration and filming.

## MORAY (TV-1A)   LAUNCHED   1964
BUILDER   U.S. Naval Ordnance Test Station, China Lake, Calif.
OWNER/OPERATOR   U.S. Navy

| | | | |
|---|---|---|---|
| NUMBER OF DIVES | 30 | PAYLOAD | 250 lbs. |
| TOTAL DIVE HOURS | 45 | LIFE SUPPORT | 72 hrs. max. |
| MAX. OPERATING DEPTH | 2,000 ft. | PILOTS 2 | |
| MAX. DEPTH ACHIEVED | 600 ft. | VIEWPORTS | none |
| COLLAPSE DEPTH | 3,000 ft. | MANIPULATORS | none |
| HULL SIZE | 2 spheres, 60 in. | PROPULSION | (1) 50-hp DC motor |
| LENGTH 33 ft. BEAM | 5.3 ft. | POWER | silver zinc battery |
| HEIGHT 5.3 ft. WEIGHT | 35,000 lbs. | CRUISE SPEED | 6 knots |
| MATERIAL | aluminum | MAX. SPEED | 16 knots |

MAJOR EQUIPMENT   No view ports, forward scan sonar, TV, UQC
STATUS   In storage; not active since 1966.

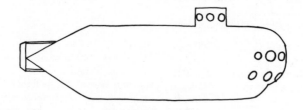

## NAIA (PC-5)   LAUNCHED   1968
BUILDER   Perry Submarine Builders, Riviera Beach, Fla.
OWNER/OPERATOR   Pacific Submersibles, Honolulu, Hawaii

| | | | |
|---|---|---|---|
| NUMBER OF DIVES | 300 | PAYLOAD | 1,000 lbs. |
| TOTAL DIVE HOURS | n. a. | LIFE SUPPORT | 36 hrs. max. |
| MAX. OPERATING DEPTH | 1,200 ft. | PILOTS 1 | OBSERVERS 2 |
| MAX. DEPTH ACHIEVED | 1,200 ft. | VIEWPORTS 13 | |
| COLLAPSE DEPTH | 2,000 ft. | MANIPULATORS | 1—general purpose |
| HULL SIZE | 54 in. | PROPULSION | (1) 7.5-hp DC motor |
| LENGTH 25 ft. BEAM | 4.6 ft. | POWER | lead acid battery 14 kwh. |
| HEIGHT 7 ft. WEIGHT | 12,000 lbs. | CRUISE SPEED | 1 knot for 8 hrs. |
| MATERIAL | mild steel | MAX. SPEED | 5 knots for 1 hr. |

MAJOT EQUIPMENT  Gyro, magnetic compass, echo sounder, underwater telephone

ESTIMATED COST  $100,000

STATUS  Completed a series of dives for pipeline inspection in Gulf of Suez, 1970. In storage.

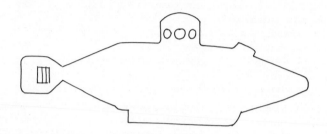

## NEKTON ALPHA    LAUNCHED  September 1968

BUILDER  Nekton, Inc.

OWNER/OPERATOR  General Oceanographics, Inc., Irvine, Calif.

| | | | |
|---|---|---|---|
| NUMBER OF DIVES | 621 | LIFE SUPPORT | 48 hrs. max. |
| TOTAL DIVE HOURS | 604 | PILOTS 1 | OBSERVERS 1 |
| MAX. OPERATING DEPTH | 1,000 ft. | VIEWPORTS (20) 11 in hull, 9 in | |
| MAX. DEPTH ACHIEVED | 1,040 ft. | tower | |
| COLLAPSE DEPTH | 2,500 ft. | MANIPULATORS 1 mechanical— | |
| HULL SIZE | 42 in. | manual | |
| LENGTH 15 ft. BEAM | 5 ft. | PROPULSION (1) 3.5-hp DC motor | |
| HEIGHT 6 ft. WEIGHT | 4.500 lbs. | POWER lead acid battery 4.5 kwh. | |
| MATERIAL | steel | CRUISE SPEED 1.5 knots for 3.5 hrs. | |
| PAYLOAD | 540 lbs. | MAX. SPEED 2.5 knots for 1 hr. | |

MAJOR EQUIPMENT  Gyro, UQC, diver-held sonar, sample chamber allows sample inspection, external lights

ESTIMATED COST  $30,000

STATUS  Diving occasionally for oil companies; other surveys.

## NEKTON BETA    LAUNCHED  September 1970

BUILDER  Nekton, Inc.

OWNER/OPERATOR  General Oceanographics, Inc., Irvine, Calif.

| | | | |
|---|---|---|---|
| NUMBER OF DIVES | 90 | LENGTH 15.5 ft. BEAM | 5 ft. |
| TOTAL DIVE HOURS | 60 | HEIGHT 6 ft. WEIGHT | 4,700 lbs. |
| MAX. OPERATING DEPTH | 1,000 ft. | MATERIAL | steel |
| MAX. DEPTH ACHIEVED | 1,040 ft. | PAYLOAD | 300 lbs. |
| COLLAPSE DEPTH | 2,500 ft. | LIFE SUPPORT | 48 hrs. max. |
| HULL SIZE | 42 in. | PILOTS 1 | OBSERVERS 1 |

# Appendices

<table>
<tr><td>VIEWPORTS</td><td>(20) 11 in hull, 9 in tower</td><td>PROPULSION</td><td>(1) 3.5-hp DC motor</td></tr>
</table>

VIEWPORTS  (20) 11 in hull, 9 in tower

MANIPULATORS  1 mechanical—manual

PROPULSION  (1) 3.5-hp DC motor

POWER  lead acid battery 4.5 kwh.

CRUISE SPEED  1.5 knots for 3.5 hrs.

MAX. SPEED  2.5 knots for 1 hr.

MAJOR EQUIPMENT  Gyro, UQC, diver-held sonar, sample chamber allows sample inspection, external lights, rock drill

ESTIMATED COST  $70,000

STATUS  Diving occasionally for oil companies; other surveys.

### NEKTON GAMMA

LAUNCHED  September 1971

BUILDER  Nekton, Inc.

OWNER/OPERATOR  General Oceanographics, Inc., Irvine, Calif.

| | | | |
|---|---|---|---|
| NUMBER OF DIVES | 10 | LIFE SUPPORT | 48 hrs. max. |
| TOTAL DIVE HOURS | 20 | PILOTS 1 | OBSERVERS 1 |
| MAX. OPERATING DEPTH | 1,000 ft. | VIEWPORTS | (20) 11 in hull, 9 in tower |
| MAX. DEPTH ACHIEVED | 1,000 ft. | | |
| COLLAPSE DEPTH | 2,500 ft. | MANIPULATORS | 1 mechanical—manual |
| HULL SIZE | 42 in. | | |
| LENGTH 15.5 ft. BEAM | 5 ft. | PROPULSION | (1) 3.5-hp DC motor |
| HEIGHT 6 ft. WEIGHT | 4,700 lbs. | POWER | lead acid battery 4.5 kwh. |
| MATERIAL | steel | CRUISE SPEED | 1.5 knots for 3.5 hrs. |
| PAYLOAD | 540 lbs. | MAX. SPEED | 2.5 knots for 1 hr. |

MAJOR EQUIPMENT  Same as on *Nekton Alpha* and *Beta*

ESTIMATED COST  n. a.

STATUS  Available for lease, as sister vehicles.

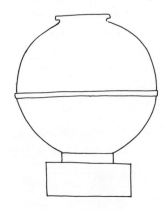

**NEMO**

# Appendices

**NEMO**  LAUNCHED 1970
BUILDER  U.S. Naval Civil Engineering Lab., Port Hueneme, Calif.
OWNER/OPERATOR  U.S. Navy

| | | | |
|---|---|---|---|
| NUMBER OF DIVES | 25 | LIFE SUPPORT | 16 hrs. max. |
| TOTAL DIVE HOURS | n. a. | PILOTS 1 | OBSERVERS 1 |
| MAX. OPERATING DEPTH | 600 ft. | VIEWPORTS | clear plastic hull |
| MAX. DEPTH ACHIEVED | 600 ft. | PROPULSION | 2 rotational motors |
| COLLAPSE DEPTH | 1,800 ft. | POWER | lead acid battery 15 kwh. |
| HULL SIZE | 66 in. | CRUISE SPEED | 1 knot for 8 hrs. |
| LENGTH | 6 ft. | MAX. SPEED | 1 knot for 8 hrs. |
| MATERIAL | acrylic | | |

MAJOR EQUIPMENT  UQC and UHF lights
ESTIMATED COST  n. a.
STATUS  Experimental use.

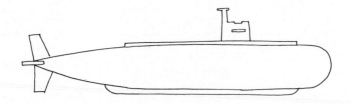

**NR-1 (NUCLEAR RESEARCH)**  LAUNCHED January 1969
BUILDER  General Dynamics/Electric Boat Div., Groton, Conn.
OWNER/OPERATOR  U.S. Navy/Sub Flot Two

| | | | |
|---|---|---|---|
| NUMBER OF DIVES | classified | LIFE SUPPORT | 45 days |
| TOTAL DIVE HOURS | classified | PILOTS 5 | OBSERVERS 2 |
| MAX. OPERATING DEPTH | classified | VIEWPORTS | several, forward |
| MAX. DEPTH ACHIEVED | classified | MANIPULATORS | 1 electrohydraulic |
| COLLAPSE DEPTH | classified | PROPULSION | 2 stern props; 4 thrusters |
| HULL SIZE | 144 in. | | |
| LENGTH 140 ft. BEAM | 12 ft. | POWER | nuclear reactor |
| HEIGHT 25 ft. WEIGHT | 400 tons | CRUISE SPEED | 5 knots for unlimited hrs. |
| MATERIAL | steel | | |
| PAYLOAD | 2,000 lbs. | MAX. SPEED | 15 knots |

MAJOR EQUIPMENT  Gyrocompass, Doppler sonar, ship control computer, CTFM sonar, TV cameras, external lights, equipped with wheels for bottom traverse photography
ESTIMATED COST  $100 million
STATUS  Detailed missions and activities—classified.

# Appendices

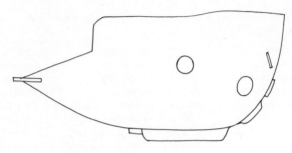

## PAULO I    LAUNCHED 1968
BUILDER   Anodics, San Diego, Calif.
OWNER/OPERATOR   Same

| | | | |
|---|---|---|---|
| NUMBER OF DIVES | 5 | LIFE SUPPORT | 48 hrs. max. |
| MAX. OPERATING DEPTH | 600 ft. | PILOTS 1 | OBSERVERS 1 |
| MAX. DEPTH ACHIEVED | 125 ft. | VIEWPORTS 6 | |
| HULL SIZE | 44 in. | MANIPULATORS | none |
| LENGTH 13.5 ft.   BEAM | 4.5 ft. | PROPULSION | (1) 4-hp DC motor |
| HEIGHT 6.3 ft.   WEIGHT | 5,200 lbs. | POWER | 868 a-h. |
| MATERIAL | steel | CRUISE SPEED | .75 knots for 2.5 hrs. |
| PAYLOAD | 400 lbs. | | |

STATUS   Never operational; new owner refurbishing for commercial diving.

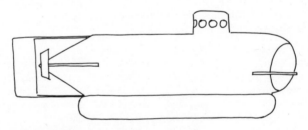

## PC-8B    LAUNCHED May 1971
BUILDER   Perry Submarine Builders, Riviera Beach, Fla.
OWNER/OPERATOR   Same

| | | | | |
|---|---|---|---|---|
| NUMBER OF DIVES | 75 | HULL SIZE | | 42 in. |
| TOTAL DIVE HOURS | 180 | LENGTH 18.5 ft.   BEAM | | 6 ft. |
| MAX. OPERATING DEPTH | 600 ft. | HEIGHT 7 ft.   WEIGHT | | 11,000 lbs. |
| MAX. DEPTH ACHIEVED | 600 ft. | MATERIAL | | steel |
| COLLAPSE DEPTH | 1,200 ft. | PAYLOAD | | 750 lbs. |

# Appendices

| | | | |
|---|---|---|---|
| LIFE SUPPORT | 48 hrs. max. | PROPULSION | (1) 7-hp DC motor |
| PILOTS 1 | OBSERVERS 1 | | stern |
| VIEWPORTS | forward hemisphere of | POWER | lead acid battery 26 kwh. |
| | glass, 120° view | CRUISE SPEED | 2 knots for 8 hrs. |
| MANIPULATORS 9 | | MAX. SPEED | 3.5 knots for 2 hrs. |

MAJOR EQUIPMENT   Magnetic compass, underwater telephone, CB radio, exploration package can be added, quick change battery pods

ESTIMATED COST   $100,000

STATUS   Dived for Canadian fisheries in Gulf of St. Lawrence, 1971.

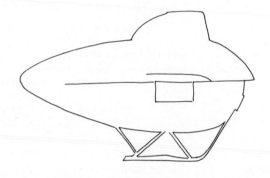

**PISCES I**   LAUNCHED   November 1965

BUILDER   International Hydrodynamics Co., Ltd., Vancouver, B.C.

OWNER/OPERATOR   Same

| | | | |
|---|---|---|---|
| NUMBER OF DIVES | 625 | LIFE SUPPORT | 68 hrs. max. |
| TOTAL DIVE HOURS | 3,400 | PILOTS 1 | OBSERVERS 2 |
| MAX. OPERATING DEPTH | 1,600 ft. | VIEWPORTS | 3 forward/down—6-in. |
| MAX. DEPTH ACHIEVED | 1,600 ft. | | diameter |
| COLLAPSE DEPTH | 3,600 ft. | MANIPULATORS | 2—salvage/recov- |
| HULL SIZE | 78 in. | | ery/sample |
| LENGTH 17 ft.   BEAM | 11.5 ft. | PROPULSION | (2) 6-hp DC motors |
| HEIGHT 9.5 ft.   WEIGHT | 14,000 lbs. | POWER | lead acid battery 60 kwh. |
| MATERIAL | Algoma 44 steel | CRUISE SPEED | 1 knot for 8 hrs. |
| PAYLOAD | 450 lbs. | MAX. SPEED | 2 knots for 4 hrs. |

MAJOR EQUIPMENT   Gyro, scanning sonar, underwater telephone, TV, still cameras, 2,000-watt external lights, UHF radio

ESTIMATED COST   n. a.

STATUS   Diving occasionally.

# Appendices

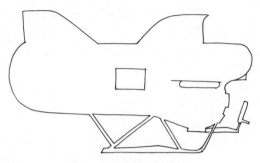

**PISCES II**     LAUNCHED  September 1968
BUILDER  International Hydrodynamics Co., Ltd., Vancouver, B.C.
OWNER/OPERATOR  Vickers Engineering, Barrow-in-Furness, England

| | | | |
|---|---|---|---|
| NUMBER OF DIVES | 300 | LIFE SUPPORT | 68 hrs. max. |
| TOTAL DIVE HOURS | 1,500 | PILOTS 1 | OBSERVERS 2 |
| MAX. OPERATING DEPTH | 3,500 ft. | VIEWPORTS | 3 forward/down—6-in. |
| MAX. DEPTH ACHIEVED | 2,400 ft. | | diameter |
| COLLAPSE DEPTH | 5,000 ft. | MANIPULATORS | 2—salvage/recov- |
| HULL SIZE | 80 in. | | ery/sample |
| LENGTH 19.5 ft. BEAM | 10 ft. | PROPULSION | (2) 6-hp DC motors |
| HEIGHT 10.3 ft. WEIGHT | 22,000 lbs. | POWER | lead acid battery 42 kwh. |
| MATERIAL | coreten B steel | CRUISE SPEED | 1 knot for 6 hrs. |
| PAYLOAD | 500 lbs. | MAX. SPEED | 2 knots for 3 hrs. |

MAJOR EQUIPMENT  Gyro, scanning sonar, underwater telephone, TV, still
        cameras, 2,000-watt external lights, UHF radio
ESTIMATED COST  n. a.
STATUS  In use with Vickers support ship.


**PISCES III**     LAUNCHED  May 1969
BUILDER  International Hydrodynamics Co., Ltd., Vancouver, B.C.
OWNER/OPERATOR  Same

| | | | |
|---|---|---|---|
| NUMBER OF DIVES | 350 | LIFE SUPPORT | 68 hrs. max. |
| TOTAL DIVE HOURS | n. a. | PILOTS 1 | OBSERVERS 2 |
| MAX. OPERATING DEPTH | 3,500 ft. | VIEWPORTS | 3 forward/down—6-in. |
| MAX. DEPTH ACHIEVED | 3,600 ft. | | diameter |
| COLLAPSE DEPTH | 5,000 ft. | MANIPULATORS | 2—salvage/recov- |
| HULL SIZE | 80 in. | | ery/sample |
| LENGTH 19.5 ft. BEAM | 10 ft. | PROPULSION | (2) 6-hp DC motors |
| HEIGHT 10.3 ft. WEIGHT | 21,000 lbs. | POWER | lead acid battery 42 kwh. |
| MATERIAL | coreten B steel | CRUISE SPEED | 1 knot for 6 hrs. |
| PAYLOAD | 500 lbs. | MAX. SPEED | 2 knots for 3 hrs. |

MAJOR EQUIPMENT  Gyro, scanning sonar, underwater telephone, TV, still
cameras, 2,000-watt external lights, UHF radio

ESTIMATED COST  n. a.

STATUS  Completed series of dives in Hudson Bay, 1970; used with support ship, *Hudson Handler*, for Aquitane Co.

## PISCES IV      LAUNCHED  Fall 1971

BUILDER  International Hydrodynamics Co., Ltd., Vancouver, B.C.

OWNER/OPERATOR  Same

| | | | |
|---|---|---|---|
| NUMBER OF DIVES | trials | PILOTS  1 | OBSERVERS  2 |
| MAX. OPERATING DEPTH | 6,600 ft. | VIEWPORTS  3 forward/down—6-in. | |
| COLLAPSE DEPTH | 9,000 ft. | diameter | |
| HULL SIZE | 80 in. | MANIPULATORS  2—salvage/recovery/sample | |
| LENGTH 19.5 ft.  BEAM | 10 ft. | | |
| HEIGHT 10.3 ft.  WEIGHT | 21,000 lbs. | PROPULSION  (2) 6-hp DC motors | |
| MATERIAL | HY-100 steel | POWER  lead acid battery 60 kwh. | |
| PAYLOAD | 1,200 lbs. | CRUISE SPEED  1 knot for 8 hrs. | |
| LIFE SUPPORT | 108 hrs. max. | MAX. SPEED  2 knots for 4 hrs. | |

MAJOR EQUIPMENT  Gyro, scanning sonar, underwater telephone, TV, still
cameras, 2,000-watt external lights, UHF radio

ESTIMATED COST  n. a.

STATUS  Delivery to USSR, February 1972.

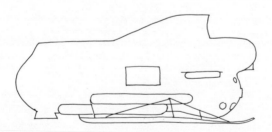

## SDL-1 (SUBMERSIBLE DIVER LOCK-OUT)
LAUNCHED  September 1970

BUILDER  International Hydrodynamics Co., Ltd., Vancouver, B.C.

OWNER/OPERATOR  Canadian Armed Forces, Fleet Diving Unit,
Halifax, N.S.

| | | | |
|---|---|---|---|
| NUMBER OF DIVES | 60 | COLLAPSE DEPTH | 4,000 ft. |
| TOTAL DIVE HOURS | 300 | (lock-out) 1,000 ft. | |
| MAX. OPERATING DEPTH | 2,000 ft. | HULL SIZE | 84 in. |
| MAX. DEPTH ACHIEVED | 2,030 ft. | LENGTH  25 ft.  BEAM | 12 ft. |

| | | | | |
|---|---|---|---|---|
| HEIGHT | 8 ft. | WEIGHT | 28,600 lbs. | MANIPULATORS 2—salvage/recovery/sample |
| MATERIAL | | | HY-100 steel | |
| PAYLOAD | | | 3,100 lbs. | PROPULSION (2) 6-hp DC motors |
| LIFE SUPPORT | | | 204 hrs. max. | POWER lead acid battery 68 kwh. |
| PILOTS 1 | | OBSERVERS | 5 | CRUISE SPEED 1 knot for 8 hrs. |
| VIEWPORTS (11) 8 forward, 2 down, 1 hatch | | | | MAX. SPEED 2 knots for 4 hrs. |

MAJOR EQUIPMENT  Gyro, scanning sonar, underwater telephone, UHF radio, TV, still cameras, 2,000-watt external lights

ESTIMATED COST  n. a.

STATUS  In use by Canadian Navy.

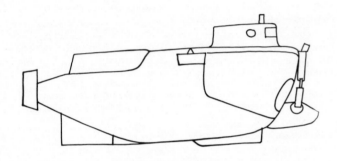

**SEA CLIFF**    LAUNCHED  December 1968

BUILDER  General Dynamics/Electric Boat Div., Groton, Conn.

OWNER/OPERATOR  U.S. Navy/COMSUBDEVGRU ONE, San Diego, Calif.

| | | | |
|---|---|---|---|
| NUMBER OF DIVES | 45 | LIFE SUPPORT | 107 hrs. max. |
| TOTAL DIVE HOURS | 85 | PILOTS 2 | OBSERVERS 1 |
| MAX. OPERATING DEPTH | 6,500 ft. | VIEWPORTS 5 | |
| MAX. DEPTH ACHIEVED | 6,500 ft. | MANIPULATORS | 2 electrohydraulic |
| COLLAPSE DEPTH | 9,750 ft. | PROPULSION | (2) 4-hp DC motors, |
| HULL SIZE | 84 in. | | (1) 5-hp stern |
| LENGTH 26 ft. BEAM | 12 ft. | POWER | lead acid battery 55 kwh. |
| HEIGHT 12 ft. WEIGHT | 48,000 lbs. | CRUISE SPEED | 1 knot for 8 hrs. |
| MATERIAL | HY-100 steel | MAX. SPEED | 2.5 knots for 1 hr. |
| PAYLOAD | 700 lbs. | | |

MAJOR EQUIPMENT  Gyrocompass, echo sounder, TV, still cameras, UQC, tool compartment for manipulators, CTFM sonar, emerging ejection capsule

ESTIMATED COST  $4 million

STATUS  Began operational diving, 1971; operating from 165-ft. ship in Hawaii during fall of 1971.

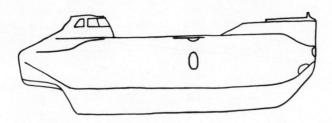

## SEVER II    UNDER CONSTRUCTION
BUILDER  USSR
OWNER/OPERATOR  USSR

| | | | |
|---|---|---|---|
| MAX. OPERATING DEPTH | 6,000 ft. | MANIPULATORS | 2 electrohydraulic |
| COLLAPSE DEPTH | n. a. | PROPULSION | 2 stern propellers; live |
| LENGTH 33 ft.  BEAM | 6 ft. | | rudder; 1 vertical |
| WEIGHT | 32,000 lbs. | POWER | lead acid battery |
| LIFE SUPPORT | 248 hrs. max. | CRUISE SPEED | 2 knots |
| PILOTS 1 | OBSERVERS 2 | MAX. SPEED | 5 knots |

MAJOR EQUIPMENT  Gyrocompass, electromechanical log, sonar, external lights, emergency relapse of ballast at 6,600 ft., light and temperature sensors

ESTIMATED COST  n. a.
STATUS  Unmanned trials to 6,600 ft. in 1970.

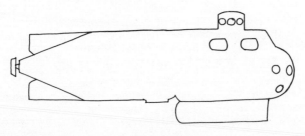

## SHELF DIVER (PLC-4)    LAUNCHED  1968
BUILDER  Perry Submarine Builders, Riviera Beach, Fla.
OWNER/OPERATOR  Perry Oceanographics Inc.

| | | | | |
|---|---|---|---|---|
| NUMBER OF DIVES | 200 | LENGTH 23 ft. | BEAM | 5.5 ft. |
| TOTAL DIVE HOURS | 600 | HEIGHT 6 ft. | WEIGHT | 18,000 lbs. |
| MAX. OPERATING DEPTH | 800 ft. | MATERIAL | | steel |
| MAX. DEPTH ACHIEVED | 800 ft. | PAYLOAD | | 2,200 lbs. |
| COLLAPSE DEPTH | 1,200 ft. | LIFE SUPPORT | | 50 hrs. max. |
| HULL SIZE | 54 in. | PILOTS 1 | OBSERVERS | 3 |

# Appendices

VIEWPORTS    (21) 8-in. diameter
MANIPULATORS    1 hydraulic; rock corer
PROPULSION    (1) 10-hp DC motor stern

POWER    lead acid battery 38 kwh.
CRUISE SPEED    1.5 knots for 12 hrs.
MAX. SPEED    3 knots for 1.5 hrs.

MAJOR EQUIPMENT    Lock-out hatch, gyro, echo sounder, underwater telephone, forward sonar, TV, video tape, exploration package of cameras and search sonar
ESTIMATED COST    n. a.
STATUS    Recently completed a survey contract for Virgin Islands government. Now in Europe for charter.

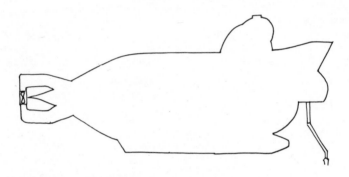

**SHINKAI (U26)**    LAUNCHED    1968
BUILDER    Kobe Shipyard of Kawasaki Heavy Industries
OWNER/OPERATOR    Japanese Maritime Safety Agency

NUMBER OF DIVES    50
TOTAL DIVE HOURS    n. a.
MAX. OPERATING DEPTH    1,960 ft.
MAX. DEPTH ACHIEVED    1,960 ft.
COLLAPSE DEPTH    n. a.
HULL SIZE    (2) 144 in.
LENGTH    46 ft.    BEAM    17 ft.
HEIGHT    15 ft.    WEIGHT    200,000 lbs.
MATERIAL    steel
PAYLOAD    6,600 lbs.

LIFE SUPPORT    n. a.
PILOTS    2    OBSERVERS    2
VIEWPORTS    4
MANIPULATORS    1—general sampling
PROPULSION    (1) 11 KW/AC stern; 2 side mounted
POWER    lead acid battery 200 kwh.
CRUISE SPEED    1.5 knots for 8 hrs.
MAX. SPEED    3.5 knots for 0.3 hrs.

MAJOR EQUIPMENT    Gyrocompass, echo sounder, underwater telephone, transponder, TV, still and cinema cameras
ESTIMATED COST    n. a.
STATUS    Began operational dives in late 1970.

232

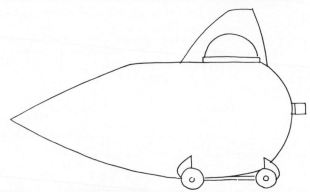

**STAR I**     LAUNCHED   1963
BUILDER   General Dynamics/Electric Boat Division, Groton, Conn.
OWNER/OPERATOR   Philadelphia Maritime Museum, Philadelphia, Pa.

| | | | |
|---|---|---|---|
| NUMBER OF DIVES | test vehicle | PAYLOAD | 200 lbs. |
| TOTAL DIVE HOURS | n. a. | LIFE SUPPORT | 18 hrs. max. |
| MAX. OPERATING DEPTH | 200 ft. | PILOTS 1 | |
| MAX. DEPTH ACHIEVED | 200 ft. | VIEWPORTS | (2) 7½-in. diameter |
| COLLAPSE DEPTH | n. a. | MANIPULATORS | none |
| HULL SIZE | 48 in. | PROPULSION | (2) ¼-hp motors |
| LENGTH 10 ft. BEAM | 6 ft. | POWER | lead acid/fuel cell |
| HEIGHT 5.8 ft. WEIGHT | 2,750 lbs. | CRUISE SPEED | 0.7 knots for 3 hrs. |
| MATERIAL | steel | MAX. SPEED | 1 knot for 1 hr. |

MAJOR EQUIPMENT   Gyro, echo sounder, CB radio, underwater telephone
ESTIMATED COST   $25,000
STATUS   On display at museum.

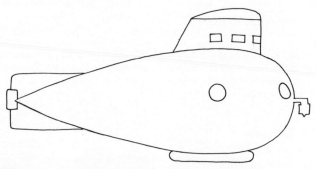

**STAR II**     LAUNCHED   1966
BUILDER   General Dynamics/Electric Boat Div., Groton, Conn.
OWNER/OPERATOR   On loan to University of Hawaii, Oahu

233

| | | | |
|---|---|---|---|
| NUMBER OF DIVES | 325 | LIFE SUPPORT | 52 hrs. max. |
| TOTAL DIVE HOURS | n. a. | PILOTS 1 | OBSERVERS 1 |
| MAX. OPERATING DEPTH | 1,200 ft. | VIEWPORTS | (6) 2 forward/down, |
| MAX. DEPTH ACHIEVED | 1,200 ft. | | 2 forward, 2 side |
| COLLAPSE DEPTH | 4,000 ft. | MANIPULATORS | 1—general purpose |
| HULL SIZE | 60 in. | PROPULSION | (2) 2-hp DC motors, |
| LENGTH 17 ft. BEAM | 5.3 ft. | | 1 vertical thrust |
| HEIGHT 7.6 ft. WEIGHT | 10,000 lbs. | POWER | lead acid battery 20 kwh. |
| MATERIAL | steel | CRUISE SPEED | 1 knot for 10 hrs. |
| PAYLOAD | 500 lbs. | MAX. SPEED | 3 knots for 1.5 hrs. |

MAJOR EQUIPMENT   Magnesyn compass, underwater telephone, CB radio,
TV, still camera, external lights

ESTIMATED COST   n. a.

STATUS   Diving at Makai Range for University of Hawaii.

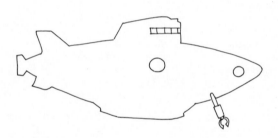

## STAR III   LAUNCHED   1966

BUILDER   General Dynamics/Electric Boat Div., Groton, Conn.

OWNER/OPERATOR   Scripps Institution of Oceanography, La Jolla, Calif.

| | | | |
|---|---|---|---|
| NUMBER OF DIVES | 250 | PILOTS 1 | OBSERVERS 1 |
| TOTAL DIVE HOURS | n. a. | VIEWPORTS | (5) 3 forward, 1 port, |
| MAX. OPERATING DEPTH | 2,000 ft. | | 1 stbd. |
| MAX. DEPTH ACHIEVED | 2,000 ft. | MANIPULATORS | 1 hydraulic |
| COLLAPSE DEPTH | 4,000 ft. | PROPULSION | (1) 7-hp stern, (1) 2- |
| HULL SIZE | 66 in. | | hp bow, (1) 2-hp ver- |
| LENGTH 24 ft. BEAM | 6.5 ft. | | tical |
| HEIGHT 9 ft. WEIGHT | 19,400 lbs. | POWER | lead acid battery 29 kwh. |
| MATERIAL | HY-100 steel | CRUISE SPEED | 1 knot for 12 hrs. |
| PAYLOAD | 1,000 lbs. | MAX. SPEED | 4 knots for 1.5 hrs. |
| LIFE SUPPORT | 122 hrs. max. | | |

MAJOR EQUIPMENT  Magnesyn compass, echo sounder, underwater tele-
phone, CB radio, TV, still camera, external lighting
ESTIMATED COST  $65,000
STATUS  Donated to Scripps, 1970. In storage.

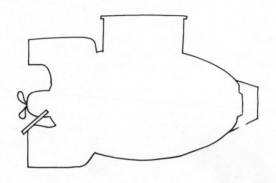

**SUBMANAUT**     LAUNCHED  1965
BUILDER  Helle Engineering, Inc., San Diego, Calif.
OWNER/OPERATOR  Same

| | | | | |
|---|---|---|---|---|
| NUMBER OF DIVES | 65 | PAYLOAD | | 1,000 lbs. |
| TOTAL DIVE HOURS | 65 | LIFE SUPPORT | | 11 hrs. max. |
| MAX. OPERATING DEPTH | 200 ft. | PILOTS  1 | OBSERVERS | 1 |
| MAX. DEPTH ACHIEVED | 350 ft. | VIEWPORTS  2 | | |
| COLLAPSE DEPTH | 2,000 ft. | MANIPULATORS | | none |
| HULL SIZE | 50 in. | PROPULSION | | 1.5-hp. DC motor |
| LENGTH  10 ft.  BEAM | 6 ft. | POWER | lead acid battery 5 kwh. | |
| HEIGHT  5.5 ft.  WEIGHT  5,000 lbs. | | CRUISE SPEED | 1 knot for 5 hrs. | |
| MATERIAL | wood and resin | MAX. SPEED | 2 knots for 2 hrs. | |

MAJOR EQUIPMENT  Compass, underwater telephone, speedometer, ther-
mometer, tracking pinger, transponder, tape recorder
ESTIMATED COST  $20,000
STATUS  Not in use.

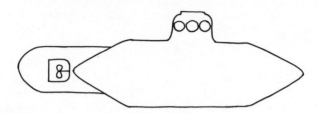

## SUBMARAY    LAUNCHED   1964
BUILDER   Douglas Privett, C & D Tools, Calif.
OWNER/OPERATOR   Kinautics, Inc., Winchester, Mass.

| | | | |
|---|---|---|---|
| NUMBER OF DIVES | 300 | LIFE SUPPORT | 32 hrs. max. |
| MAX. OPERATING DEPTH | 300 ft. | PILOTS 1 | OBSERVERS 1 |
| MAX. DEPTH ACHIEVED | 325 ft. | VIEWPORTS 14 | |
| COLLAPSE DEPTH | 1,000 ft. | MANIPULATORS 1 mechanical— | |
| HULL SIZE | 36 in. | manual | |
| LENGTH 12.5 ft. BEAM | 3 ft. | PROPULSION (1) 35-hp DC motor | |
| HEIGHT 5 ft. WEIGHT | 2,900 lbs. | POWER lead acid battery 10 kwh. | |
| MATERIAL | steel | CRUISE SPEED 2 knots for 8 hrs. | |
| PAYLOAD | 380 lbs. | MAX. SPEED 3 knots for 4 hrs. | |

MAJOR EQUIPMENT   Gyro, forward scan sonar, CB radio, external lights, underwater telephone
ESTIMATED COST   $30,000
STATUS   Idle, in storage.

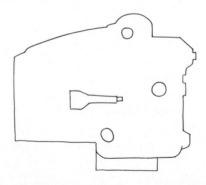

## SURV (STANDARD UNDERWATER RESEARCH VEHICLE)
LAUNCHED   1968
BUILDER   Lintott Engineering Ltd., Foundry Lane, Horsham, England
OWNER/OPERATOR   Same

236

| | | | |
|---|---|---|---|
| NUMBER OF DIVES | 10 | PILOTS 1 | OBSERVERS 1 |
| MAX. OPERATING DEPTH | 1,000 ft. | VIEWPORTS 4 | |
| MAX. DEPTH ACHIEVED | 600 ft. | MANIPULATORS 2—for sampling | |
| HULL SIZE | 63 in. | and instruments | |
| LENGTH 10 ft. BEAM | 7 ft. | PROPULSION (2) 4-hp AC motors, | |
| HEIGHT 8.8 ft. WEIGHT | 10,200 lbs. | port and stbd. | |
| MATERIAL | steel | POWER lead acid battery 12 kwh. | |
| PAYLOAD | 250 lbs. | CRUISE SPEED 1 knot for 4 hrs. | |
| LIFE SUPPORT | 50 hrs. max. | MAX. SPEED 2.5 knots for 1 hr. | |

MAJOR EQUIPMENT  Gyrosyn compass, echo sounder, underwater telephone, surface radio

ESTIMATED COST  n. a.

STATUS  Inoperative and in storage.

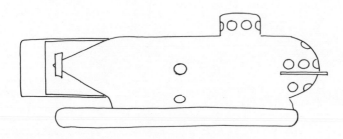

## SURVEY SUB (PC-9C)    LAUNCHED  June 1970
BUILDER  Perry Submarine Builders, Riviera Beach, Fla.
OWNER/OPERATOR  Brown & Root Engineering Corp., Houston, Tex.

| | | | |
|---|---|---|---|
| NUMBER OF DIVES | 100 | LIFE SUPPORT | 100 hrs. max. |
| MAX. OPERATING DEPTH | 1,200 ft. | PILOTS 1 | OBSERVERS 2 |
| MAX. DEPTH ACHIEVED | 1,350 ft. | VIEWPORTS (21) 8-in. diameter | |
| COLLAPSE DEPTH | 3,000 ft. | MANIPULATORS none | |
| HULL SIZE | 56 in. | PROPULSION (1) 10-hp DC motor, | |
| LENGTH 22 ft. BEAM | 8 ft. | 1 thruster | |
| HEIGHT 7 ft. WEIGHT | 22,000 lbs. | POWER lead acid battery 40 kwh. | |
| MATERIAL | steel | CRUISE SPEED 1–1.5 knots for 18 hrs. | |
| PAYLOAD | 1,200 lbs. | MAX. SPEED 4.5 knots for 1.5 hrs. | |

MAJOR EQUIPMENT  Doppler navigator, object-locating sonar, TV, video tape, cameras, and lights

ESTIMATED COST  $250,000

STATUS  In use by Brown and Root Co. for survey purposes and pipeline inspection. Diving in mid-East.

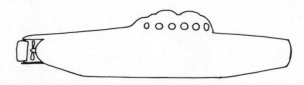

**TECHDIVER (PC-3B)**   LAUNCHED  1964
BUILDER   Perry Submarine Builders, Riviera Beach, Fla.
OWNER/OPERATOR   International Underwater Contractors, Flushing, N.Y.

| | | | |
|---|---|---|---|
| NUMBER OF DIVES | 700 | PAYLOAD | 750 lbs. |
| TOTAL DIVE HOURS | 2,000 | LIFE SUPPORT | 20 hrs. max. |
| MAX. OPERATING DEPTH | 600 ft. | PILOTS  1 | OBSERVERS  1 |
| MAX. DEPTH ACHIEVED | 600 ft. | VIEWPORTS  13 | |
| COLLAPSE DEPTH | 1,000 ft. | MANIPULATORS | 1—general purpose |
| HULL SIZE | 48 in. | PROPULSION | (1) 7-hp DC motor |
| LENGTH  20 ft.  BEAM | 3.5 ft. | POWER | cable for surface power |
| HEIGHT  5 ft.  WEIGHT | 6,500 lbs. | CRUISE SPEED | 2 knots for 8 hrs. |
| MATERIAL | steel | MAX. SPEED | 4 knots for 1 hr. |

MAJOR EQUIPMENT   Magnetic compass, echo sounder, CB radio, under-
water telephone
ESTIMATED COST   $70,000
STATUS   Available for diving.

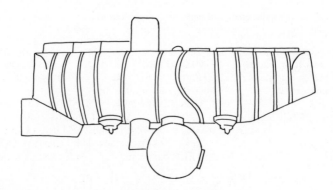

**TRIESTE I**   LAUNCHED  1953
BUILDER   Auguste Piccard and City of Trieste, Italy
OWNER/OPERATOR   U.S. Navy/USNEL, San Diego, 1958–1964

# Appendices

| | | | |
|---|---|---|---|
| NUMBER OF DIVES | 128 | PAYLOAD | 5,000–24,000 lbs. |
| TOTAL DIVE HOURS | 600 | LIFE SUPPORT | 48 hrs. max. |
| MAX. OPERATING DEPTH | 36,000 ft. | PILOTS 1 | OBSERVERS 1 |
| MAX. DEPTH ACHIEVED | 35,800 ft. | VIEWPORTS (2) 1 forward, 1 aft | |
| COLLAPSE DEPTH | 45,000 ft. | MANIPULATORS | 1 electric |
| HULL SIZE | 74 in. | PROPULSION 2 stern, 1 bow thruster | |
| LENGTH 58 ft. BEAM | 11.5 ft. | POWER silver zinc battery 20 kwh. | |
| HEIGHT 26 ft. WEIGHT 317,000 lbs. | | CRUISE SPEED | 0.6 knots for 4 hrs. |
| MATERIAL | steel | MAX. SPEED | 0.8 knots for 3 hrs. |

MAJOR EQUIPMENT  Compass, UQC, altitude/depth sonar, still camera, lights

ESTIMATED COST  $150,000

STATUS  Retired after *Thresher* search in 1964. Float on display at U.S. Navy Yard, Washington, D.C.

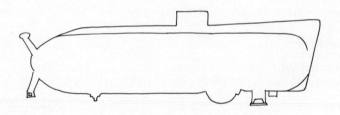

**TRIESTE II**    LAUNCHED  July 1965
BUILDER  U.S. Navy, Mare Island Naval Shipyard, San Francisco, Calif.
OWNER/OPERATOR  U.S. Navy/COMSUBDEVGRU ONE, San Diego, Calif.

| | | | |
|---|---|---|---|
| NUMBER OF DIVES | 80 | LIFE SUPPORT | 45 hrs. max. |
| TOTAL DIVE HOURS | n. a. | PILOTS 1 | OBSERVERS 2 |
| MAX. OPERATING DEPTH | 20,000 ft. | VIEWPORTS | 1 forward/down |
| MAX. DEPTH ACHIEVED | 13,100 ft. | MANIPULATORS | 2 electrohydraulic |
| COLLAPSE DEPTH | 40,000 ft. | PROPULSION (2) 10-hp, (1) 2-hp | |
| HULL SIZE | 84 in. | bow thruster | |
| LENGTH 67 ft. BEAM | 15 ft. | POWER silver zinc battery 145 kwh. | |
| HEIGHT 3 ft. WEIGHT | 50 tons | CRUISE SPEED | 2 knots for 5 hrs. |
| MATERIAL | steel | MAX. SPEED | 3 knots for 2 hrs. |
| PAYLOAD | 20,000 lbs. | | |

MAJOR EQUIPMENT  Gyrocompass, Doppler sonar, transponder navigator, altitude/depth sonar, TV, UQC, UHF radio, still cameras, and lights

ESTIMATED COST  n. a.

STATUS  Modified for *Scorpion* search, 1969. Now diving off San Diego.

# Appendices

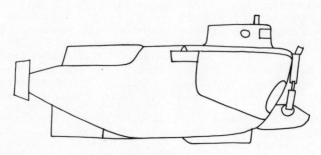

**TURTLE**  LAUNCHED  December 1968
BUILDER  General Dynamics/Electric Boat Div., Groton, Conn.
OWNER/OPERATOR  U.S. Navy/COMSUBDEVGRU ONE, San Diego, Calif.

| | | | |
|---|---|---|---|
| NUMBER OF DIVES | 20 | LIFE SUPPORT | 107 hrs. max. |
| MAX. OPERATING DEPTH | 6,500 ft. | PILOTS 2 | OBSERVERS 1 |
| MAX. DEPTH ACHIEVED | 6,500 ft. | VIEWPORTS 5 | |
| COLLAPSE DEPTH | 9,750 ft. | MANIPULATORS | 2 electrohydraulic |
| HULL SIZE | 84 in. | PROPULSION (2) | 4-hp DC motors, |
| LENGTH 26 ft. BEAM | 12 ft. | (1) 1-hp stern | |
| HEIGHT 12 ft. WEIGHT | 48,000 lbs. | POWER | lead acid battery 55 kwh. |
| MATERIAL | HY-100 steel | CRUISE SPEED | 1 knot for 8 hrs. |
| PAYLOAD | 700 lbs. | MAX. SPEED | 2.5 knots for 1 hr. |

MAJOR EQUIPMENT  Gyrocompass, echo sounder, TV, still cameras, UQC, tool compartment for manipulator, CTFM sonar, emergency ejection capsule
ESTIMATED COST  $4 million
STATUS  Began operational diving in 1971 with sister ship, *Sea Cliff*, from support ship, *Maxine*, in San Diego.

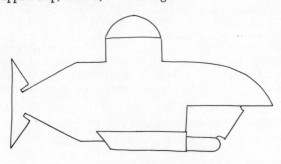

**VAST MARK III**

**VAST MARK III**     LAUNCHED  1967
BUILDER   Underwater Vehicles, Inc., Waterford, Conn.
OWNER/OPERATOR   UVI owns several; approximately 12 *Mark III* built
since 1967

| | | | |
|---|---|---|---|
| NUMBER OF DIVES | 1,000 | LIFE SUPPORT | 1.5 hrs. max. |
| TOTAL DIVE HOURS | no record | PILOTS 1 | |
| MAX. OPERATING DEPTH | 250 ft. | VIEWPORTS | (1)  16-in.  forward/ |
| MAX. DEPTH ACHIEVED | 250 ft. | | down, 1 hatch |
| COLLAPSE DEPTH | 825 ft. | MANIPULATORS | 1 mechanical |
| HULL SIZE | 36 in. | PROPULSION | (2) ½-hp DC motors |
| LENGTH  10 ft.    BEAM | 4.7 ft. | POWER | lead acid battery 3 kwh. |
| HEIGHT   5 ft.   WEIGHT | 2,500 lbs. | CRUISE SPEED | 1 knot for 6 hrs. |
| MATERIAL | steel | MAX. SPEED | 2 knots for 6 hrs. |
| PAYLOAD | 280 lbs. | | |

MAJOR EQUIPMENT   Magnesyn compass, lights, underwater telephone
ESTIMATED COST   $15,000
STATUS   Completed contract for Navy Underwater Center in Caribbean in
1971. Demonstration dives. Other *Mark III* vehicles owned by
Brown and Root Corp., Houston, Texas, and National Underwater
Contractors, Canada.

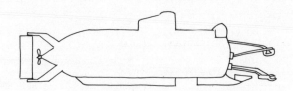

**YOMIURI**     LAUNCHED  1964
BUILDER   Kobe Shipyards, Mitsubishi Heavy Industries, Kobe, Japan
OWNER/OPERATOR   Yomiuri Shimbun Newspaper, Tokyo, Japan

| | | | | |
|---|---|---|---|---|
| NUMBER OF DIVES | 435 | PILOTS 2 | | OBSERVERS  4 |
| TOTAL DIVE HOURS | 1,324 | VIEWPORTS | (7) 3 in hull, 4 in | |
| MAX. OPERATING DEPTH | 1,000 ft. | | tower | |
| MAX. DEPTH ACHIEVED | 1,000 ft. | MANIPULATORS | 1 general purpose, | |
| COLLAPSE DEPTH | 1,900 ft. | | sample | |
| HULL SIZE | 78 in. | PROPULSION | (1) 12 KW/1,000 rpm | |
| LENGTH  48 ft.    BEAM | 8.2 ft. | | motor | |
| HEIGHT 12.4 ft.   WEIGHT | 74,000 lbs. | POWER | diesel/lead acid 450 a-h. | |
| MATERIAL | steel | CRUISE SPEED | 2 knots for 10 hrs. | |
| PAYLOAD | 6,500 lbs. | MAX. SPEED | 3 knots for  6 hrs. | |
| LIFE SUPPORT | 492 hrs. max. | | | |

# Appendices

MAJOR EQUIPMENT  Compass, underwater telephone, still and movie cameras, external lights, forward scan sonar, echo sounder, current meter

ESTIMATED COST  n. a.

STATUS  Continued use in shallow water surveys around Japanese Islands.

Several other submersibles exist which have insufficient information available for separate listing.

| Name | Length (ft.) | Depth Operation (ft.) | Owner | Remarks |
|---|---|---|---|---|
| Viperfish | 20 | 1000 | Dan Taylor | Dived in Loch Ness |
| Snooper | 12 | 1000 | Sea Graphics | Photography work |
| Martin U-9 | 7 | 300 | Marine Resources, Inc. | |
| Mini Diver (MD-1) | 16 | 250 | Great Lake Underwater Sports | |
| Nautilette | 12 | 100 | Nautilette, Inc. | |
| Sea Ray (SRD-101) | 21 | 1000 | Submarine Research & Development | |
| Deep Six | 16 | 150 | Deep Six Marine Service | |

# B.   *Alvin*[*] Pre-Dive Inspection Check List

[*]With permission Alvin Operations Group, Woods Hole Oceanographic Institution.

a.   Conning Tower
  (1)   Check 3 sail vent holes clear.
  (2)   Check that upper hatch latches tight.
  (3)   Check 3 windows for looseness and cracks.
  (4)   Main Ballast System
      (a)   Check hand vent valve opens and then shut.
      (b)   Check for air leaks through vent line.
      (c)   Check Main ballast box for oil leaks.
      (d)   Check oil level in box by testing tension in diaphragm.
      (e)   Check piping for loose fittings.
  (5)   Check SP phone jack plugged.
  (6)   Check all electrical leads not in use, if any, plugged.
  (7)   Hatch
      (a)   Inspect "O" ring and "O" ring groove.
      (b)   Grease "O" ring.
      (c)   Check outside jacking screw by turning.
      (d)   Inspect hatch seat for surface defects.
      (e)   Inspect hatch window for looseness and marks inside and outside.
      (f)   Close and dog hatch. Check fully seated all around. Leave open.
      (g)   Check dogs retracted.
      (h)   Put dog wrench in clip.
  (8)   Check CT Drain valves free, and clear of debris.
  (9)   Remove hatch cover and hatch seat cover from C.T.
b.   Inside Sphere
  (1)   Check hatch dog wrench in place.
  (2)   Check emergency drill and bit in place.
  (3)   Hull release
      (a)   Using SR wrench, turn ⅛ turn right, then ⅛ turn left.
      (b)   Inspect for loose fittings.
      (c)   Check aligned with mark.
      (d)   Replace wrench in clip.
  (4)   Penetrators
      (a)   Using P. wrench, check all 12 nuts for tightness.
      (b)   Check spacer properly oriented.

243

# Appendices

       (c)   Replace wrench in clip.

(5)   Oxygen System
       (a)   Check pressure 1800 psi.
       (b)   Open stop valve and check flow.
       (c)   Shut stop valve.
       (d)   Check $O_2$ monitor in place and calibrate.

(6)   (a)   Zero Altimeter.
       (b)   Zero barometer.

(7)   $CO_2$ Removal System
       (a)   Install fresh LiOH canister.
       (b)   Install new charcoal filter.
       (c)   Check blower running and shut off.
       (d)   Check $CO_2$ monitor in place and calibrate.

(8)   Check emergency battery voltage. Reading _____.

(9)   Check Control Circuit Switch on
       (a)   "Normal"
       (b)   Emergency
       (c)   Science

(10)   Check Gyro running and settled out.

(11)   Check magnetic compass runs and turn off.

(12)   SCUBA                  1_____
       (a)   Check pressure on bottles. Reading:   2_____
                                     3_____
                                     4_____
       (b)   Open stop valve and check 2 regulators. Leave *open*.
       (c)   Check 2_____ masks in place.
              3_____

(13)   Check cabin lighting.

(14)   Check instrument lighting.

(15)   Check inverter(s) running.

(16)   Check mounting frames tight.

(17)   Check for loose gear. Stow securely.

(18)   Check 4 windows for defects and properly seated.

(19)   Check Battery Condition meters and record.
       _____V Propulsion Bus
       _____V Control Bus
       _____V Science Battery.

(20)   Check Variable ballast sphere pressure and record.
       _____psi.

(21)   Check List Angle.   Record)
                                ) When waterborne

(22)   Check Trim Angle.   Record)

(23)   Main Propulsion
       Note: *Have person monitoring from topside to keep persons*
              *clear.*
       (a)   Variable volume pump       Amps. _____
       (b)   Variable volume pump       Amps. _____

244

# Appendices

  (c) Fixed volume pump    Amps. _____
  (d) Rudder right and left. Check indicator.
  (e) Lift propellers up - down - check indicators and align.
    Amps _____
  (f) Main propeller.
  (g) Life propellers - ahead; astern; P ahead Stbd astern;
           S ahead P astern.
  (h) Secure System
(24) Variable Ballast (Check main valve opens and shuts)
  (a) Pump 100 lbs.
  (b) Flood 100 lbs.
    Check pressure indication.
  (c) Compensate for installed payload.
  (d) Check valve shut.
  (e) Secure Systems.
(25) Trim System
  (a) Run pump fwd for 15 secs.  Amps _____
  (b) Run pump aft for 15 secs.  Amps _____
  (c) Check trim angle if in water. Adjust to zero or as directed.
  (d) Secure sytem.
(26) Main Ballast System
  (a) Check manual vent shut.
  (b) Operate vents, observe indication open and shut.
  (c) Operate blow for ten seconds.
    (1) Observe valve opens and shuts
    (2) Operate list control port and starboard
  (d) Check vents and blow valve shut.
  (e) Secure system.
(27) Check underwater television Amps _____
  Operation and secure.
(28) Check underwater telephone Amps _____and secure.
(29) Check SP phone in place and plugged in.
(30) Check 2 diver flashlights working and in place.
(31) Check radio and secure.
(32) *If in water,* check echo sounder and secure. Amps_____
(33) Check Doppler Navigation System and secure.
(34) In water, check Leak Detectors for no indication.

c. LOWER HULL—WALK AROUND—STARTING AT BOW
 (1) *Sonar Pocket*
  (a) Sonar in      Secure.
  (b) Sonar out
    (1) Drain holes   Clear.
    (2) Cover     On/Off.
 (2) *Forward Window*
  (a) Scratches, crazing
  (b) Retaining rings   Secure.
  (c) Plexiglas shields  On/Off

# Appendices

      (d)  M/P drain                                Clear

      (e)  Window insert                           Fastened

| | | |
|---|---|---|
| (d) | M/P drain | Clear |
| (e) | Window insert | Fastened |
| (f) | Window insert removed | |
| | (1) Electric penetrators | Properly installed and safety wired. |
| | (2) Hull paint | Scratches, cracks. |
| (3) | *Manipulator* | |
| (a) | **On** | |
| | (1) Stowed | |
| | (2) Electrical leads | Properly connected |
| | (3) Joints | Grease-filled. |
| (4) | *Lower Window* | |
| (a) | Scratches and crazing | |
| (b) | Retaining rings | Secured |
| (c) | Plexiglas shields | On/Off |
| (d) | Mercury drain | Clear |
| (e) | Window insert | Fastened |
| (f) | Window insert removed | |
| | (1) Hull alignment | Proper |
| | (2) Emergency release | Engaged |
| | (3) Electric disconnects | Seated, compensated and no leaks. |
| | (4) V/B plumbing | Proper |
| | (5) V/B Pressure transducer | Installed and connected. |
| | (6) Level of mercury | Checked |
| | (7) Explosive valves | Connected |
| | (8) Electric Penetrators | Properly installed and safety wired. |
| | (9) Wires | Stowage, connections/caps. |
| | (10) General | Hydraulic leaks, hull or hardware corrosion |
| (5) | *V/B system box drain* | Off, no leaks |
| (6) | *Battery oil reservoir drain* | Off, no leaks |
| (7) | *Forebody Buoyancy Package—* Ballast tank fastening | Secure. |
| (8) | *Port and Starboard windows* | |
| (a) | Scratches and crazing | |
| (b) | Retaining rings | Secure. |
| (c) | Plexiglas shields | On/Off |
| (9) | Ballast Tanks and Sponsons | Damage or cracks |
| (10) | Forebody-afterbody couplings | Engaged |
| (11) | Port fathometer transducer | Inspected |
| (12) | Upper Afterbody | Securely fastened |
| (13) | V/B reservoir cover fastenings (port and starboard) | Secure |
| (14) | Main propulsion box | Securely fastened |

246

| | | |
|---|---|---|
| (a) | Covers on | |
| | (1) Hose and fittings | Secure |
| | (2) Leaks | None |
| (15) | Battery boxes | Clear to release |
| (16) | Aft mercury drain | Clear |
| (17) | U/W Telephone transducer | Secure |
| (18) | Stern post dome | Leaks, cracks |
| (19) | Stern, shroud | Chips, cracks |
| | (a) free motion within limits | |
| | (b) propulsion hydraulic fittings | No leaks |
| (20) | Stern propeller | Undamaged, free turning |

d. *UPPER HULL AND CONNING TOWER EXTERIOR*

| | | |
|---|---|---|
| (1) | Forward and side windows | Scratches and fastenings |
| (2) | Tower vent | Clear |
| (3) | Sponsons fastenings | Secure |
| (4) | Port and starboard lift motors | Aligned |
| | (A) Propellers | Undamaged, free |
| (5) | Battery charging cover | |
| | (a) Removed | |
| | (1) Electric leads | Capped |
| | (2) Main propulsion and mercury trim fill lines | Capped |
| | (3) Battery fill lines one foot oil head | Checked |
| | (b) Cover | Secured |
| (6) | After padeye | Secured |
| (7) | Radio antenna | Secure |
| (8) | Water speed indicator | Elevated |

## INSPECTION RECORD SHEET

Date _____

a) *Servicing and Inspection*

The submarine has been serviced and inspected. _____

(Signature)

| *Discrepancies* | *Action* | *Initial* |
|---|---|---|

# Appendices

b) *Pilot Acceptance and Dive Data*

Dive # _____

Location _____                              Date _____

Passengers _____                          Time _____

Having reviewed previous sheets and the servicing record
I accept ☐      down ☐      the submarine.      _____

(Pilot's Signature)

c)


*Dive Record and Discrepancies*            Max. Dive Depth _____
                                                         Length of Dive _____


| Discrepancies | Action | Date | Initial |
|---|---|---|---|
|  |  |  |  |
|  |  |  |  |

Submarine is up ☐      down ☐      _____

(Pilot's Signature)

248

# C. Institutions, Corporations, and Agencies Using Submersibles

Chicago Bridge and Iron Corporation, 901 West 22nd Street, Oakbrook, Illinois.

Florida Atlantic University, Department of Ocean Engineering, Boca Raton, Florida. Professor Charles R. Stefan, Chairman.

General Dynamics/Electric Boat Division, Groton, Connecticut.

General Motors, A.C. Division, 6767 Hollister Avenue, Goleta, California.

General Oceanographics-Nekton Inc., 17911 Bascom Street, Irvine, California.

Grumman Aircraft Engineering Corporation, Bethpage, Long Island, New York.

International Hydrodynamics, 624-736 Granville Street, Vancouver, British Columbia.

International Underwater Contractors, 33-25 127th Street, Flushing, New York.

Kinautics Inc., 1 Lowell Street, Winchester, Massachusetts.

Lockheed Missiles and Space Corporation, Sunnyvale, California.

National Marine Fisheries Service, Washington, D.C.

North American Rockwell Corporation, Ocean Systems Operations, Seal Beach, California.

Perry Oceanographics, Submarine Builders, Riviera Beach, Florida.

Reynolds Submarine Service, 615 SW 2nd Avenue, Miami, Florida.

Smithsonian Institution, Washington, D.C.

Sun Shipbuilding and Dry Docks Corporation, Chester, Pennsylvania.

Texas A&M University, Department of Oceanography, College Station, Texas. Dr. Richard A. Geyer, Department Head.

Underwater Vehicles, Inc., P.O. Box 242, Waterford, Connecticut.

U.S. Navy Civil Engineering Laboratory, Port Hueneme, California.

U.S. Navy Oceanographic Office, Suitland, Maryland.

U.S. Navy Ships System Command, Washington, D.C.

U.S. Navy SUBLANT, Groton, Connecticut.

# Appendices

U.S. Navy Submarine Development Group One, San Diego, California.

U.S. Navy Underwater Sound Laboratory, New London, Connecticut.

University of California, Scripps Institution of Oceanography, La Jolla, California. Dr. Edward L. Winterer, Chairman and Professor of Geology.

University of Hawaii, Department of Oceanography and Ocean Engineering, Honolulu, Hawaii. Dr. Charles L. Bretschneider.

University of Miami, Rosenstiel School of Marine and Atmospheric Sciences, Miami, Florida. Dr. F. G. Walton Smith, Director.

University of Rhode Island, Graduate School of Oceanography, Kingston, Rhode Island. Dr. John A. Knauss, Dean.

University of Southern California, Department of Geological Sciences, Los Angeles, California. Dr. O. L. Bandy, Professor of Geological Sciences.

University of Texas, Marine Science Institute and Defense Research Laboratory, Port Aransas, Texas, Dr. Donald E. Wohlschlag, Director.

Westinghouse Ocean Research and Engineering Center, Box 1488, Annapolis, Maryland.

Woods Hole Oceanographic Institution, Department of Biology, Chemistry, Geology, Geophysics, and Physical Oceanography. Dr. H. Burr Steinbach, Dean, Graduate School.

# Glossary

*AC.* Alternating current. An electrical current that reverses direction continuously at regular intervals, usually 60 cycles per second. AC is produced by an inverter that takes the DC from the batteries and makes alternating current. Propulsion motors driven by AC are more efficient, lighter, and more reliable than those driven by DC.

*Air Lock.* A small chamber with an outer hatch and an inner hatch that allows submersible personnel to "swim out" by equalizing pressure in the chamber with ambient sea pressure. *Deep Diver* and *Beaver* are examples of boats that have an air lock or swim-out hatch.

*Altitude/Depth Sonar.* A type of sonar that measures the distance from the submersible to the ocean surface or to the ocean bottom, using upward and downward transducers and a graphic recorder.

*Ballast.* Weight in the form of water, lead, iron pigs, or shot, used to change the displacement of a submersible.

*BG.* A naval architectural term for the distance in inches between the center of gravity and the center of buoyancy. A submersible must have a positive BG to be stable. Smaller vehicles of five to ten tons require a BG of one or two inches, while larger vehicles require large BG for sufficient stability.

*Bladder.* A flexible bag that usually contains a liquid used to fill a submarine battery and equalize the sea pressure.

*Brow.* A lightweight framework designed to allow the mounting of sensors such as lights, cameras, etc., several feet beyond the bow of a vehicle. Additional syntactic foam may be carried on the brow for increasing the payload. Brows usually can be jettisoned in emergency.

# Glossary

*Buoyancy.* The upward force exerted by a fluid upon a submersible. Buoyancy can be achieved by air contained in tanks or by materials that are lighter than seawater, such as gasoline or syntactic foam.

*Camera obscura.* A darkened chamber in which the real image of an object is received through a small opening or lens and projected in natural color onto a facing surface.

*Connector.* A plug and socket for easy connecting and disconnecting of exterior instruments and sensors, such as cameras and lights. Connectors are usually mated when the submersible is out of the water and dry. One type, called "self-purging," can be mated underway and keeps moisture from getting on the contacts.

*Core.* A vertical, cylindrical sample of bottom sediments that allows a geologist to determine the nature and stratification of the bottom. Several vehicles can take a short sediment core, and a few can rotary core hard rock.

*Corer.* A metal or plastic tube that can be driven into the bottom sediments either hydraulically or by tilting the vehicle. Some corers can take more than one core.

*CTFM Sonar.* Continuous Transmission Frequency Modulation is a type of sonar that sends out a continuous sonic beam that is reflected off hard objects ahead of the submersible. A scanner hydrophone picks up the sound and presents a picture, showing distances to objects such as rocks, ships' hulls, etc.

*CURVE.* Controlled Underwater Research Vehicle. An unmanned, tethered vehicle operated by NUC for salvage and recovery to 7,000 feet.

*DC.* Direct current. An electrical current that flows in one direction, such as that eminating from a storage battery used in a submersible. DC is used for propulsion in the majority of submersibles, since no conversion or conditioning is required, making it less expensive.

*Directional Gyro.* Also known as a free gyro, this type of compass has a drift of several degrees per hour, as well as the rotation of the earth, amounting to a relatively inaccurate way of determining heading for a submersible.

*Dry Boat.* A submersible that operates at one-atmosphere (approximately 14.7 psi) by use of a pressure hull, and is dry inside.

# Glossary

*DSL.* Deep Scattering Layer. A stratified population of organisms in the ocean which reflect or scatter sound. Layers or concentrations of animals are found at depths of 100 to 400 fathoms. There are usually several types of organisms. Some migrate diurnally to the surface.

*DSRVG.* Deep Submergence Review Vehicle Group. Created by the U.S. Navy in 1963 to investigate the loss of the submarine *Thresher.*

*DSSP.* Deep Submergence Systems Program of the U.S. Navy that resulted from DSRVG in 1968. The office was combined with Naval Ship Systems Command in 1970.

*Fish.* A towed body, usually carrying instruments.

*Fjarlie bottle.* A series of metal tubular bottles mounted outside a submersible that allow water to be sampled by remote-control valves. Used on *Deepstar.*

*Gyrocompass.* A gyroscopically-driven compass that shows a true North orientation rather than magnetic north. Gyrocompasses are relatively drift free and precise, but are heavy, expensive, and require moderate power for the smaller submersibles.

*Hardwire Phone.* A communication link used at the surface between the support ship and submersible during launching. The wire or cable is disconnected just before the vehicle is released. All of the tethered vehicles have a hardwire phone for continuous communication.

*Hard Tank.* A tank designed to withstand maximum submergence pressure, used to hold water or oil that is pumped out into a rubber bladder to change displacement. Part of the variable ballast system.

*Hosing.* A term applied to cables which means that water or air can travel inside the cable jacket. All submergence cables should be of a nonhosing type to prevent leaks and corrosion.

*Hydrophone.* A type of transducer that receives sonar energy from the water and converts it to electrical energy for sonars and underwater communication. Hydrophones may be arranged in arrays, such as a line mounted on a boom.

*Hyperbaric operation.* Pressurized activities in excess of two atmospheres in dry/wet boats that have a swim-out capability.

# Glossary

*KWH.*  Kilowatt hours. The amount of watts in thousands used per hour.

*Magnesyn Compass.*  An electric compass that uses a magnetic reference for a heading. Not of comparable accuracy with a gyrocompass. Used on smaller submersibles.

*NAVOCEANO.*  Naval Oceanographic Office located in Washington, D.C. which has a mission for deep ocean survey.

*NEL.*  Navy Electronics Laboratory. Now NUC, San Diego, Calif.

*NUC.*  U.S. Naval Underwater Center, San Diego, Calif.

*NURDC.*  U.S. Naval Research and Development Center, San Diego, Calif.

*NUSL.*  U.S. Navy Underwater Sound Laboratory, New London, Conn.

*OBSS.*  Ocean Bottom Scanning Sonar. One of the first side-scan sonars developed by Westinghouse Underseas Division. This sonar sends narrow beams of sonic energy to either side of a towed body or a submersible, and receives a reflected pattern back that shows a map of the terrain comparable to an aerial photograph.

*Outgassing.*  Battery outgassing occurs during charging and to a lesser extent during discharge, and requires proper venting of gases. Lead acid batteries give off hydrogen gas which is highly explosive, and for this reason the batteries are nearly always carried externally and pressure compensated and vented with a one-way valve.

*Payload.*  The amount of weight a vehicle can carry, excluding crew, observers and their personal effects. Usually, payload is made up of scientific instruments which are not a regular part of the boat's instruments.

*Penetrator.*  A passageway for wires that goes through the pressure hull, allowing the wires to pass but restricting water. One penetrator may carry a number of wires or leads which carry power or signals.

*Pinger.*  An acoustic device that emits a sonic signal at a particular frequency and acts as a locating device for a submerged submersible.

*Pitch.*  Motion in the fore-and-aft plane, such that the bow dips and rises.

# Glossary

*PQC.* Navy designation for a diver-held sonar used to locate objects undersea by sound.

*Roll.* Motion about the longitudinal axis, from side to side.

*Rotary Inverter.* An instrument that electronically changes DC power to AC, using a rotating device.

*Scrubber.* The part of the life support system, such as lithium hydroxide or baralyme and associated fans, trays, and ducting, that removes carbon dioxide.

*Self-purge.* A type of connector that cleans or wipes all moisture from the contacts as it is being mated. It may be used underwater where sensors have to be removed without bringing the vehicle out of the water.

*SIO.* Scripps Institution of Oceanography, La Jolla, Calif.

*Slurp gun.* A plastic syringe type of gun used to "slurp" in small fish. A modification for submersibles uses an electrically driven propeller to suck water into a funnel as a vacuum cleaner would.

*Static Inverter.* An instrument that electronically changes DC power to AC, using solid state electronic components.

*STD.* Salinity-Temperature-Depth. An instrument developed to measure continuously the variables of salinity, temperature, and depth which are important to physical oceanography. One such instrument was mounted on *Alvin*.

*Sub-bottom profiler.* An acoustic instrument carried on submersibles that emits a high energy sonar pulse of high frequency which penetrates the upper layers of sediment. The hydrophone receives the returned signal and a recorder shows the geologic strata of the upper 50 to 100 feet.

*Submarine.* Any manned vehicle that is able to operate independently of the surface or needs no surface support for handling and supplies, and can operate under its own power on the surface as well as submerged.

*Submersible.* Any manned vehicle that depends upon a surface support tender or ship either for towing or shipboard carrying as well as resupply, recharging and life support materials.

# Glossary

*Syntactic foam.*   Composed of phenolic or glass microballoons within an epoxy matrix. This foam is used for supplementary buoyancy. Densities of foam vary from 30 to 42 pounds per cubic foot. The lighter foams with greater lift capacity tend to absorb water at greater depths.

*Thruster.*   A propeller or water jet, usually located at the bow and stern, used to give lateral thrust to the vehicle to aid in slow speed steering or maneuvering. Thrusters may be oriented vertically and horizontally. Some are ducted, which channels the flow of water in a tube or duct, and makes the thruster more efficient.

*Toroidal Submarine.*   A proposed large nuclear submarine with a hollow center section used to launch, retrieve, and house smaller D/RVs.

*Transducer.*   A device such as a piezoelectric crystal that converts electrical energy input into a sonic output into the water, or the converse. Sonars and underwater telephones use transducers outside the submersible which usually are oil-filled for pressure compensation.

*Transponder.*   A sonar device that listens for a signal to trigger a transmitter that replies with a sonic output, usually on a separate frequency. Transponders are used in bottom-mounted or vehicle-mounted navigation systems.

*Trim system.*   Any of several variable ballast or weight systems are used to change trim fore and aft. A commonly used system is to pump mercury from an after tank to a forward tank hydraulically. A simple system is to move batteries or other weights fore and aft.

*UQC.*   Navy designation for the wireless, underwater "telephone" which uses sonic energy in the water to communicate between the surface support ship and a submerged submersible. UQC has been nicknamed "Gertrude" by the Navy and is usually a frequency around 8KHz.

*Variable Ballast System.*   Any of several types of pumping systems that can change ballast by admitting seawater or expelling it, by pumping oil from a hard tank to a soft bladder, or by admitting water to tanks and dropping weights to allow the vehicle to dive or rise.

256

# Glossary

*Varivec Propeller.* A type of multibladed propeller that has a variable pitch control that allows high maneuverability when used on a submersible. First developed by Westinghouse Electric Corp.

*Velocimeter.* An instrument that measures the speed of sound in seawater.

*Wet Boat.* A free flooding, ambient pressure submersible, designed to be operated by swimmers using some sort of breathing apparatus, either air or mixed gas. Wet means that the inside is flooded, as opposed to a dry boat.

*WHOI.* Woods Hole Oceanographic Institution, Woods Hole, Mass.

*Xenon Flasher.* A light that emits a bright flash and uses a strobe tube and fresnel lens for the purpose of surface identification. These lights are usually visible for two miles at night.

*Yaw.* Motion about the vertical axis.

# Bibliography

Ballard, Robert D., and Emery, K. O. *Research Submersibles in Ocean-ography*. Washington, D.C.: Marine Technology Society, 1970.

Barton, Otis. *World Beneath the Sea*. New York: The Thomas Y. Crowell Co., 1953.

Beebe, William. *Half-Mile Down*. New York: Duell, Sloan and Pearce, 1951.

Behrman, Daniel. *The New World of the Oceans*. New York: Little Brown, 1969.

Cohen, Paul. *The Realm of the Submarine*. Toronto: The Macmillan Co., 1970.

Cousteau, Jacques-Yves, and Dugan, James. *The Living Sea*. New York: Harper & Row, 1963.

Cousteau, Jacques-Yves, and Dugan, James. *World Without Sun*. New York: Harper & Row, 1964.

Cox, Donald W. *Explorers of the Deep*. Maple, N. J.: Hammond Inc., 1968.

Cross, Wilbur. *Challengers of the Deeps*. New York: Wm. Sloan Assoc., 1959.

Davis, Robert H. *Deep Diving and Submarine Operations*. London: Siebe Gorman Co., 1962.

Dugan, James. *Man Under the Sea*. New York: Collier Books, revised ed. 1965.

Emery, K. O.; Ballard, Robert; and Wigley, Roland. "A Dive Aboard *Ben Franklin* Off West Palm Beach, Florida," *Marine Technology Society Journal*, March, 1970, pp. 7–16.

*Guidelines for the Selection, Training and Qualifications of Deep Submersibles Pilots*. Deep Submersible Pilots Association, 1967.

Houot, G. S., and Willm, P. H. *2,000 Fathoms Down*. New York: E. P. Dutton & Co., 1955.

Interagency Committee on Oceanography. *Deep Research Vehicles*. Washington, D.C.: ICO, 1966. (Publication 18)

# Bibliography

Lake, Simon P. *The Autobiography of Simon Lake, as told to Herbert Grey*. New York: Appleton-Century-Crofts, 1938.

*Manned Certification Procedures Criteria Manual for Manned Non-Combatant Submersibles*. Washington, D.C.: United States Naval Ship Systems Command, 1968.

Markel, Arthur. "What Has the Submersible Accomplished?" *Oceanology International*, July, 1968, pp. 34–38.

Martin, George W. *Trieste, the First Ten Years*. United States Naval Institute Proceedings, August, 1964.

Morris, Richard K. *John P. Holland, 1841-1914, Inventor of the Modern Submarine*. Annapolis, Maryland. United States Naval Institute, 1966.

Neumann, A. Conrad. "The Submersible as a Scientific Instrument." *Oceanology International*, July, 1968, pp. 39–42.

Piccard, Auguste. *Earth, Sky, and Sea*. New York: Oxford University Press, 1956.

Piccard, Jacques, and Dietz, Robert S. *Seven Miles Down*. New York: G. P. Putnam's Sons, 1960.

Rainnie, William O. "Recovery of *DSRV Alvin*," *Ocean Industry*, IV/11 (November 1969), pp. 61–63, IV/12 (December, 1969), pp. 69–70.

*Safety and Operational Guidelines for Undersea Vehicles*. Washington, D.C.: Marine Technology Society, 1968.

Shenton, Edward H. *Exploring the Ocean Depths*. New York: W. W. Norton & Co., Inc., 1968.

Shenton, Edward H. "Where Have All the Submersibles Gone?" *Oceans Magazine*, November, 1970, pp. 39–56.

Shepard, Francis P., and Dill, Robert. *Submarine Canyons and Other Sea Valleys*. Chicago: Rand McNally, 1966.

Sweeney, James B. *A Pictorial History of Oceanographic Submersibles*. New York: Crown Publishers, Inc., 1970.

Terry, Richard D. *The Deep Submersible*. North Hollywood, Calif.: Western Periodicals Co., 1966.

Thomas, Lowell. *Sir Hubert Wilkins: His World of Adventure*. New York: McGraw Hill, 1961.

Wilkins, Sir George Hubert. *Under the North Pole: The Wilkins-Ellsworth Submarine Expedition*. New York: Brewer, Warren & Putnam, 1931.

# Index

## A

Academy of Science (USSR), 194
Acoustics, underwater, 129–30
*Albacore,* 37
*Aluminaut,* 75–76, 77, 84, 86, 131, 132, 136, 141, 161, 162, 179, 193
  *Alvin* salvage operation, 151–53; average dive length, 153; diving accomplishments of, 147, 148–53; H-bomb search mission, 142–43, 148–49; manipulators, 113–14; power system, 93, 94; preliminary sea trials, 146; sonar equipment, 148–49; surface support for, 118–19; wheels, 150
Aluminum, use of, 86–87
*Alvin,* 75–76, 87, 92, 100, 107–8, 111, 131, 136, 141, 142–43, 147, 159, 162, 179, 193
  ballast system, 101–2; buoyancy, 114–15; collision with *Lulu,* 145, 152; control features, 105–6; coring tools, 110; coring tubes, 109; diving capacity, 142; emergency devices, 116; fish attack on, 114; geologic explorations, 127; H-bomb search mission, 142–43; naval certification of, 168; number of dives, 146; propulsion system, 98–99; refurbished, 143–44; search and salvage of, 132, 151–53; sinking of, 146; S.T.D., 129; support/landing system, 117
*Alvin II,* 187–88
*Alvin*-type vehicles, 188

American Bureau of Shipping (ABS), 167, 168–69, 170
American Revolution, 23, 24
American Submarine Company, 28–29
Anti-submarine warfare (ASW), 130, 133, 183
Aqualungs, 70
*Archimede,* 56, 110, 126, 154, 190–91
*Argonaut I,* 38–39, 42, 150
*Argonaut Jr.,* 38
*Argyronete,* 192
*Army and Navy Journal,* 29
*Asherah,* 77
Atlantic Underwater Test and Evaluation Center (AUTEC), 187–90
*Auguste Piccard,* 131
*Autec I,* 187–88

## B

Ball, Dr. Mahlon, 149
Ballard, Robert, 140
Ballast systems, 100–103
  conventional submarine, 101
Barham, Eric, 124–25, 157
Barton, Otis, 16, 43, 44, 47, 49, 51, 53, 56, 82, 84, 121
Bathyscaphe, meaning of, 51
Battery systems, 92–95
Bauer, Wilheim, 27
Beagles, John, 155
*Beaver IV,* 111, 116, 133, 189
Beebe, (Charles) William, 16, 43–50, 51, 52, 53, 56, 61–62, 82, 121

261

# Index

262

# Index

# Index

# Index

# Index

# Index